Praise for *Our Angry Eden: Faith and Hope on a Hotter, Harsher Planet*

"A beautiful and timely reminder that there's plenty we can be doing to fit in better with creation. Read it and change!"
—Bill McKibben, founder of 350.org
and author of many books, including *Falter* and *Eaarth*

"When it comes to abusing creation, we have the science, knowledge, and understanding that we need to do the right thing, but we lack the spiritual fortitude. In *Our Angry Eden*, David Williams tends to our courage by giving voice to our spiritual crisis, longing, and hope."
—Carol Howard Merritt,
writer and pastor of Bedford Presbyterian Church

"We have been waiting for more Christians to speak out and speak up about the dangers of our climate emergency. David Williams is doing just this in an engaging manner. This book is personal, prophetic, and practical."
—Mary Evelyn Tucker, cofounder and codirector of the Forum on Religion and Ecology at Yale University; coauthor of *Journey of the Universe*

"*Our Angry Eden* is a must-read. With clear-eyed honesty and a perceptive analysis of the existential threat of the climate crisis, Williams forces us to face the mess we are in. But he also conjures hope through lively storytelling, biblical insight galore, and sound practical ideas that embody God's good future. From beginning to end, the book is animated by sparkling prose. This is one of the best books I have read on what it means to follow Jesus in this tumultuous time."
—Steve Bouma-Prediger, author of
For the Beauty of the Earth: A Christian Vision for Creation Care and
Earthkeeping and Character: Exploring a Christian Ecological Virtue Ethic

"In a time when the words *climate emergency* send many people into a panicked despair or an overwhelmed paralysis, David Williams offers a third way. Williams shows how a full-fledged response to the climate crisis is a profound act of faith and integral to following in the way of Jesus. This book gives attainable, tangible ways to engage, while spreading out a rich theological foundation for how to love our neighbor as we care for our earthly home."

—Anna Woofenden, author of *This Is God's Table: Finding Church Beyond the Walls* and founder of The Garden Church

OUR
ANGRY
EDEN

OUR ANGRY EDEN

FAITH & HOPE ON A HOTTER, HARSHER PLANET

David Williams

OUR ANGRY EDEN
Faith and Hope on a Hotter, Harsher Planet

Cover design: Joel Holland

Print ISBN: 978-1-5064-7044-3
eBook ISBN: 978-1-5064-7045-0

At Sanctuary today in this fateful hour,
I place all heaven with its power,
And the sun with its brightness,
And the snow with its whiteness,
And Fire with all the strength it hath,
And Lightning with its rapid wrath,
And the winds with their swiftness along their path,
And the sea with its deepness,
And the rocks with their steepness,
And the Earth with its starkness,
All these I place
By God's almighty help and grace,
Between myself and the powers of Darkness.

ST. PATRICK, "FAEDH FIADA"

CONTENTS

PART IV

Following Jesus after Climate Change

PART I

THE WORLD OF OUR CREATION

CHAPTER 1

TAKING CARE IN GOD'S CREATION

I arrived at the church for a gathering. The theme of the gathering was creation care. *Creation care*, if you're not familiar with the term, is a theological approach to our environment, the idea that we should love and cherish the earth that God gave us. It's a lovely, sweet idea.

We Jesus-folk gathered, and we sang. Drums were beaten, because my denomination bends liberal. Good, warm words from Scripture were spoken, some of them in Hebrew, because we are Presbyterian and we enjoy that sort of thing. Earnest songs were sung in delightful harmony. At the end of the event, we were given a charge: go home and share your reflections on creation care with friends, family, neighbors, and acquaintances.

So I sat, and I thought for a while, and then I bore down and wrote this book.

There is an assumption among religious progressives—and it is a well-meaning one—that human beings need to care for creation because it is so terribly fragile. Here we are, dumping plastic into our seas and filling our shallow skies with the carboniferous flatulence of our strange, anxious busyness. We are razing our mountaintops to tear out their profitable hearts. We are raising and slaughtering creatures by the hundreds of millions in industrial facilities, inflicting horrors upon horrors on simple, suffering beings. We are reaching deep into the seas and scraping them bare, denuding them of life.

Poor creation, we think. Poor dolphins and butterflies. Poor lambs and baby polar bears, we think. We must protect our poor fragile planet, we think.

There is also an assumption among religious liberals that we should protect creation for aesthetic reasons. Because it is beautiful. I don't for a moment dispute this. Seas and stars, storms and aurora? Beautiful. Life itself, from the tiniest budding crocus to the serene majesty of a blue whale? Amazing, complex, miraculous. I am deeply sympathetic to this position.

As anyone who knows me will tell you, I'm the first one out there when severe weather rolls in. When the local news station cuts to Storm Team special coverage, I'm almost giddy. That feeling of the wind rising, of the sweet, sharp, electric tang in the air as the sky grows dark and rich and dangerously alive? I love it.

And when creation is smaller and slower and more subtle, I find it equally fascinating. When I was a little boy, I could spend an entire afternoon watching ants at work. I could spend a whole Southern summer evening spitting watermelon seeds and marveling at trees alight with the faerie glow of fireflies.

I see the stars, I hear the rolling thunder, as the old hymn goes: God's power throughout the universe displayed. When creation speaks, I pay attention. I am fascinated.

But there's another truth, another facet to how I view the world around me. God's creation scares the heck out of me. (Lord have mercy, does no one read Jack London anymore? No, I suppose y'all don't.)

The delicate balance of our planet's ecosystem to which we've spent the last few million years adapting? It can be rough but nowhere near as harsh as it is becoming. As we change the makeup of the atmosphere upon which we rely for life, and as the temperatures of our world begin their inexorable rise upward, we should expect to receive back what creation returns to those who meddle with it or imagine they rule it.

If we abuse it, it will not tolerate us. God's creation owes us nothing, and it isn't known for being forgiving.

Our angered Eden is red in tooth and claw, as Alfred, Lord Tennyson, puts it. It is as implacable as the rising sea, or the storm that scours and shatters, or the fire that rages with the driving wind. It has always been a terrifying thing. The fierce new world that we have woken is no less willing to break us than our old familiar world, and it's a whole bunch meaner.

And that's just our tiny, agitated blue speck of a planet. When I say the words "God's creation," I don't think of the earth. *Creation* is not a synonym for *earth* to me. Not at all.

I think of all of it. All thirteen-point-something billion years of our space-time, stretching gigaparsecs beyond the parochial scale of our imaginings. And beyond, past the firmament of our local physics, into a possibly multiversal infinity that goes deeper still, to terrifying deeps beyond deep.

God didn't just make this small, rocky world. You look up to the twinkling stars, so pretty in the sky? That's a great yawning vastness, filled with fire and emptiness and poison, where life is desperately rare and hangs on by a thread. In most of it, we *Homo sapiens sapiens* can survive about five seconds before we burn or freeze solid, assuming there's no explosive decompression involved.

Creation is not just our world, dagflabbit.

We need to take "care" of God's creation in the way that we take "care" when we find ourselves bobbing precariously on a surfboard with a great white shark moving lazily in the murky shadows beneath. We need to take "care" of it in the way that we take "care" when we teeter on the ledge of a precipice.

We are not good at this kind of creation care anymore. In fact, we've gotten worse at it as we've progressed technologically as a species.

Oddly enough, the humans who lived at the time the Bible was written were more than aware of the terrifying reality of God's work. The storm and the fire and the sea were terrifying. The One who made them all was to be feared even more. Life was short, and death was ever present. Premoderns were aware that they hung on by a thread, one that the Fates could cut on a whim.

But we moderns are coddled fools, wrapped in a few hundred fleeting years of industrial agriculture and fossil fuels and a false sense of our own power. Postmodernity, with its subjectivity and relativism? That hasn't helped things either, as our grasp on the very idea of the real slips away. We whisper lies to ourselves in the closed minds of our social media #delusionchambers.

Our little bit of earth does not care about our desires and fantasies at all. If we sabotage our ecosystem, this ecosystem, we might survive. I hope we do. But the ensuing tumult of five thousand years of warming might also bring us down and leave another, less maladaptive species to rise in our place. Do we think creation cares? It does not.

Creation would continue as if nothing had happened. After a hundred million years, there'd be no trace of us at all. An epochal spasm of mass extinctions that could wash us away would mean little to our rocky world—just another epoch among epochs. It would matter even less on the true scale of creation.

This is, admittedly, a grim thing to say, particularly if you're used to a faith that focuses only on personal benefits and prosperity. But what of God? Here I am, a follower of Jesus and a believer in God, and I'm supposed to be talking about hope and happiness. I mean, God does care, right?

Yes. God does. But.

God also allows us to reap the harvest we have sown, no matter what that harvest might be. That is the nature of both God's love and God's justice.

And sometimes God lets us die. In fact, God lets us die pretty much all of the time. One hundred percent of us die in the hands of God, being that we're mortal and all.

Christians seem happy to forget this. Our God, say the progressives, is loving and inclusive and just. The God of the Bible, say the conservatives, watches over the Bible-believing righteous and has a plan for us all. Our God, say the prosperity-gospel churches, will make us all rich if we just plant a seed for our pastor's new jet.

None of us seem to be looking to the Creator of the Universe and saying, "The Creator of the Universe fills me with mortal terror."

We should be.

Not because God is a monster. But because justice is the nature of being, meaning that all systems in God's creation seek equilibrium. This reality, which underlies the natural law of creation? It does not bode well for us right at the moment as we blithely, blindly traipse down a path to ruin.

So when I hear my fellow Jesus-folk singing earnestly about care for creation, and about how pretty and nice and fragile it all is, I know there is truth in that.

But that warm, easy, butterflies-and-rainbows sensibility does not mesh with the depth of the existential threat we have created for ourselves and our children. Our soft, green songs do not speak to the coming howl of the rising wind, or the searing heat of the sun, or the roar of the devouring sea. They do not sing of hardship and struggle or of the powerful, resilient faith that will be needed to survive the harsh and unforgiving world we have inflicted on ourselves.

Journalists and scientists have written a lot over the past few years that has called attention to the great transition in our climate. In 2019, a book of climate horrors entitled *The Uninhabitable Earth* rested for a while at the number one spot on the *New York Times* Best Seller list. "Doomer" writings and "apocalypse porn" are increasingly part of the story we tell ourselves about our world. Catastrophe and apocalypse are trendy things, and it's easy to misconstrue the climate crisis we're facing as just another fad or panic. The bone-deep anxiety that many folks feel now about the future of our children on a harsher world is interpreted by some as a creature of this odd, hyperaware moment in history.

But it isn't a recent thing. We've known about the impacts our emissions are having on our world for decades. More than ten years ago, environmentalist Bill McKibben wrote a book called *Eaarth*. It's a short, readable, and grim account of McKibben's struggle against our complacency and distraction in the face of a crisis. The core thesis of the book was that the time to avoid the climate crisis had passed and that our world is now compromised. "Eaarth" is the name McKibben gave to the new planet our home is becoming. Eaarth is a world that isn't just in our future but already upon us, filled with raging storms and spreading deserts, where life itself is harder. It's a world that doesn't take our crap and doesn't suffer fools lightly. It is not a world that will be bullied, ignored, or beaten into submission. It is not a world that will submit to our ideological blinders and our blind greed. It will not be a safe space for our anxious, narcissistic fragilities.

This is the new planet we are discovering, a world of our creation that will grow harsher and harsher as decades and the centuries pass. Our children know this, know that we have consigned them to a world that will increasingly not be the one we inherited. It is why young people march, and demonstrate, and try to call us to attention.

It's a new world that seems painfully familiar. As an author of modestly successful dystopian science fiction novels, I've played around with all kinds of different apocalyptic scenarios. Most of them are riffs on familiar TEOTWAWKI (The End of the World as We Know It) tropes, with aliens and zombies and robot uprisings, oh my. In my novel *When the English Fall*, a massive solar storm wipes out all electronics, leaving the Amish as the only ones standing. As I spun out that story, though, I bumped right up against a hard truth: I couldn't tell a near-future story without folding weird weather into it. It wouldn't have felt real. When real apocalypses start impinging on your fantastic tales, it means something.

I'm also a pastor. I've felt a particular calling in my years of ministry to serve small congregations where my job is, well, peculiar. In little, intimate churches, the job of the pastor isn't to be the executive or the manager. You're not the patriarch or matriarch, he or she who must be obeyed. A healthy small church is more organic, with authority distributed based on giftedness and reinforced by trust. My role is to interpret, teach, and inspire my sweet corner of the beloved community to live out the gospel every single day. Abstract theology and Big Social Issues are only relevant insofar as they're a real part of our lives together. Climate change may be big, but it's also part of our lives, right here, right now.

And as a disciple of Jesus, I can say this with certainty: the gospel has a place in this new world. Jesus will be important to those of us who will endure this coming time, and the gospel will be as vital on McKibben's Eaarth as it was on our old world.

This book calls us to recognize that reality and to name the ways the gospel speaks to us in this climate crisis. It asks us to realize how even more fiercely relevant it is to follow Jesus when things have gotten hard. We'll start out looking at where we are, right here in this section. How can we be so sure global warming is a real issue? And if it is, why haven't we just up and fixed it? We haven't, of course. From there, we'll move into looking at four reasons we've failed to slow or reverse climate change. Hint: it's because human beings are kind of a mess. That we're a mess doesn't mean we have no ground to respond to this epochal challenge. The basis for that response, for Christians, is and must be based in Christ's teachings. That moral ground is not an abstraction. Morality gives purpose, and it is the basis for framing and guiding action. Finally, we'll explore the question, What does that action look like? It looks remarkably like the traditional Christian virtues. Adaptability and hospitality. Sabbath keeping and humble living. Compassion for God's creatures and grace toward all. An openness to all gifts and a learning heart. And above all, hope, hope in the face of our anxious fears.

Let's get to it, because the time to deal with it is now.

CHAPTER 2

CLIMATE CHANGE IS REAL

We have all seen it.

In 2018, we watched North Carolina, battered by storm-driven rain, roads covered, highways closed. We watched Wilmington, cut off for days, as rivers rose and turned the city into an inaccessible island. We looked to the Florida Forgotten Coast as a small tropical storm blossomed into a nightmarish beast, functionally overnight, and as entire coastal towns were obliterated by winds and the sea.

That hurricane crossed Florida into Georgia, where it was still a hurricane. It crossed Georgia and the Carolinas, and when it entered my home state of Virginia, it was still so strong, it killed half a dozen people. It crossed a thousand miles of land, and it was still deadly.

We saw record flooding in Texas, and states of emergency were declared as the Midwest was overcome with rain. We saw the Arizona State Fair canceled at the height of summer because it was underwater.

We saw California burn, in conflagrations that have no precedent, as terrified Americans fled firestorms that roared through entire communities. As they died in their homes. As they burned to death in their cars. As they hid in their backyard pools, rising only to fill their lungs with superheated air.

These are the things that we all saw on our screens. But there is a reality beyond what we see on our magic devil boxes.

If you live in the Washington, DC, area, which I do, you saw things that people elsewhere may not have seen.

Here in Washington, DC, we had a storm in 2018. For two whole days, the wind howled—sustained winds of forty-five to fifty miles an hour, with higher gusts. Trees were down everywhere. We lost power for a day, and some of my nearby family lost power for multiple days. The storm damaged roofs everywhere, tearing away siding, pulling shingles from subroofing. Our own roof was damaged as, hour after hour, the relentless, howling wind slowly peeled the vent from our roof as I watched helplessly from our front yard. Many homes in our area weren't repaired until weeks and months afterward.

It was one of the fiercest storms I've ever seen in my lifetime living on the East Coast, made peculiar by this: it involved not a single drop of rain. No thunder. No lightning. It was, during the day, partly sunny, yet with winds that never, ever let up, leaving destruction in their wake.

We saw this, in Washington, DC, in 2018.

Here in the suburbs of Washington, DC, this year, I have a garden in my front yard. I grow green beans and kale and potatoes. I have blueberry bushes, which mostly feed the birds, and strawberries, which lately have been a favorite of the chipmunks. I've tried carrots, which have mostly not done anything at all. I seed-save my green beans and my kale, leaving pods on the healthiest plants, where they dry and provide me with next year's crops.

But in 2018, it rained. It rained endlessly, sometimes for a week straight. It was, in point of provable fact, the single wettest year in recorded history in the Washington area. Roads flooded, over and over again. The Potomac overtopped its banks. Some small towns in the DC area were obliterated by apocalyptic deluges, their streets filled with roaring, waist-high torrents. People died.

It rained so much that I lost most of the seeds I was saving—some to rot, but most of them to seeding. There in their pods, the seeds sprouted while still on the plant, the roots springing out from the still unfallen pods in what amounted to accidental hydroponics.

I shared this with other gardeners, and they concurred. It was weird. Not normal. Wrong.

I saw this, in Washington, DC, this last year.

There were other things. We saw the trees holding their leaves deep and late into a strangely delayed fall. Among the trees, the oaks were masting, wildly overproducing acorns, which they do when stressed. There were so many acorns in my backyard that they piled up in mounds.

These are the things you saw if you lived in Washington, DC, and your eyes were open. They are signs. There are so many signs, in fact, that you'd need to be the world's greatest fool not to see them.

The world is changing, right here in front of us, and the changes are everywhere and all at once. It isn't just happening nationally and locally; it's a global phenomenon. It's not just my personal observation; it's also the combined insights of science, government, and business.

Climatologists and meteorologists are unified in their assessment of this event. Their consensus is overwhelming. In a statement issued by the American Meteorological Society in 2019 summarizing the state of the science, the conclusion was inescapable: climate change is real, it's happening, and it's caused by human beings and our use of fossil fuels. We know the mechanism, we can see the culprit, and it all holds together. The theory that climate change is both happening and caused by humans is the most cohesive, logical, and testable hypothesis to explain why things are trending the way they are trending.

Beyond the professional meteorologists who observe this phenomenon, there are others ringing the alarm bell: those liberal, tree-hugging environmental activists in the US military. (This is sarcasm.) For all of the branches of the US Armed Forces, in fact, the impacts of our changing world are a strategic reality that poses a direct threat to military operations. A Department of Defense report in 2019 found that two-thirds of US strategic assets are currently facing potential threats from climatic shifts. That same report also noted that increasing impacts of climate change would have destabilizing impacts around the world, raising the likelihood of resource-driven conflicts and migration.

It's a dry, factual, and grim picture of the future, one that includes the following statement in its conclusion: "It is relevant to point out that 'future' in this analysis means only 20 years in the future. Projected changes will likely be more pronounced at the mid-century mark; vulnerability analyses to mid- and late-century would likely reveal an uptick in vulnerabilities (if adaptation strategies are not implemented.)"

This goes beyond the obvious impacts of sea level rise at US naval bases, which for some reason tend to be located in coastal areas. All branches of the military are affected in ways that aren't just theoretical or speculative. At the end of 2018, for example, the same catastrophic, explosive hurricane that devastated Florida's Panhandle took out Tyndall Air Force Base (AFB), destroying buildings and dozens of F-22 stealth fighters. A full evacuation wasn't possible, given the abrupt strengthening of the storm. Damage to the base and materiel was estimated at $4.7 billion. Massive and unprecedented flooding in early 2019 took down the US Strategic Command at Offutt AFB in Nebraska after a concerted effort by airmen to create a sandbag cordon proved inadequate.

The military takes this seriously. So do major corporations—like, say, oil companies. Not that there's any irony in that. Oil companies, in their own internal planning documents, have acknowledged the impacts of fossil fuels on the climate for decades. Both Shell and Exxon raised the alarm internally back in the 1980s, noting that the changes in the environment would have a catastrophic effect on their business operations.

With a tremendous proportion of their refining facilities positioned along the coastlines of the world, significant sea level increases and increases in storm activity pose a direct threat to fossil fuel production.

Then there's the world's largest and most widespread form of Christianity, the Catholic Church. The Catholic Church claims as members nearly half of the followers of Jesus on our little planet, with nearly 1.3 billion souls. There was a time, as Galileo would be happy to remind you, when Catholicism wasn't exactly the most science-friendly movement. This is no longer the case and hasn't been for most of the twentieth and all of the twenty-first centuries. Catholicism has a fierce and sustained intellectual tradition, one that creates a particularly analytic mindset (as can be evidenced by the fact that more than half of America's Supreme Court justices are Catholic).

As a global Christian movement, the Catholic Church also cares about what affects its members around the world, and in 2015, it called out the threat of climate change. In a papal encyclical entitled *Laudato Si'*—in the event that Latin ain't your thing, that means "Praise Be to You"—the church officially acknowledged that climate change was a problem, that the scientific basis of that statement is solid, and that human behavior and our fossil fuel–driven economy is primarily to blame.

There is no question, again, about what is happening. This is not a liberal or conservative issue, or a Democrat or Republican issue. It's reality. Remember reality? It's that thing that happens no matter what your opinion of it is.

Neither is there any question about why it's happening. We human beings are causing climate change.

The threat is clear, documented, and acknowledged by everyone who has any knowledge of the issue. That does not mean that there are not millions of human beings who resist the idea that our world's climate is changing. They are very convinced that they are correct, for reasons typically having to do with ideology. For some reason that is beyond me, American conservatism seems particularly amenable to this, and perfectly bright human beings turn their intellects and their energies toward skepticism and cynicism. They doubt the witness of science and the witness of enterprise: "Oh, that's just the liberals and big business trying to scare us into giving away our freedom." Even when the US Army and Navy raise the alarm, that cynicism rises up: "Well, you know that the brass is just using this to get more funding." I'm not quite sure how that cynicism works with the large oil companies: "Well, that's just big business covering

its own butt because, um . . . err . . . something." In the face of unprecedented storms and global heat waves, they'll shake their heads: "Oh, that's
just weather. There've always been heat waves."

I know such souls, and I love them as Christian brothers and sisters.
But their reflexive cynicism reminds me of a story from my favorite childhood storyteller: C. S. Lewis. I must have read Lewis's Narnia books a
dozen times through, and even though I've got a BA in religious studies
and a master's degree and a doctorate, I'm still convinced that I learned
everything that matters about Christian faith in the green fields and forests of Narnia.

In *The Magician's Nephew*, the creation story of the world of Narnia,
there's a character named Uncle Andrew. He's the magician of the title,
a man completely convinced of his own correctness in all things. When
he finds himself watching Aslan—the great lion who's the Jesus of that
world—singing Narnia into existence from the void, he fights against
what he sees:

> When the Lion had first begun singing, long ago when it was still quite
> dark, he had realized that the noise was a song. And he had disliked the
> song very much. It made him think and feel things that he did not want
> to think and feel. Then, when the sun rose and he saw that the singer was
> a lion ("only a lion," as he said to himself) he tried his hardest to make
> himself believe that it wasn't singing and never had been singing—only
> roaring as any lion might in a zoo in our own world. "Of course it can't
> really have been singing," he thought. "I must have imagined it. I've been
> letting my nerves get out of order. Who ever heard of a lion singing?"
> And the longer and more beautifully the Lion sang, the harder Uncle
> Andrew tried to make himself believe that he could hear nothing but
> roaring. Now the trouble about trying to make yourself stupider than
> you really are is that you very often succeed.

It's a point that C. S. Lewis makes several times in his Narnia tales:
human beings are entirely capable of convincing themselves of anything,

even if it means denying what's right in front of their noses. Just because we're bright and capable doesn't mean we can't find ways to blind ourselves to the world around us. And while being stubborn and sticking to your guns can, in some ways, be a useful way to bull your way through the world, it isn't exactly a virtue that lends itself to learning and personal growth.

Denial can also stand in the way of seeing what's right there in front of you. The brightness of your carefully constructed rationalizations dazzles your eyes and keeps you from seeing reality. Human beings are remarkably talented at this.

I've had a lot of conversations with one of my neighbors over the years. He's a grizzled Frenchman, a bright, aggressive, obstreperous person, a man whose home is a complex shambles of half-finished projects. He's a former electrical engineer, and it's clear that he once lived a much more normal life. Now, unfortunately, he is also completely insane. He is utterly convinced that the world is an impossibly complicated web of conspiracies against him—conspiracies that involve local police, politicians, major insurance companies, the French government, and an array of agents and spies who monitor him constantly wherever he goes. Every time a police car moves through the neighborhood, it's because they're watching him. Every time a new neighbor who he doesn't recognize walks past his house, they're an agent hired by the company that provides his disability insurance.

I've tried, of course, to tell him that this just ain't so. But his world is too tightly crafted around himself and around the elegantly interwoven constructs of delusion that give him a sense of control. Nothing anyone says will convince him otherwise.

"I'm not crazy," he'll say to me. "I'm just an [profanity]. Who are you, who believes in your stupid fairy tale about God, to tell me about my own life?" So he sees what he sees, everything filtered through his preconceptions, and nothing more.

Our own ideological frameworks can easily do that to us, to the point where no amount of evidence will open our eyes. Those blinders only bring us harm, both to our souls and to our persons. If, on the other hand, you see what's coming—if you're not too dazzled by ideology and your situational awareness remains intact—you're primed to act.

So if we know climate change is a threat, what can be done about it? How can we stop it?

CHAPTER 3

TOO LITTLE, TOO LATE

Our changing climate is a huge problem and an existential threat, and it might seem at first like it's impossible to solve. It's easy to look at the climate crisis, throw up your hands, and despair.

But there are plenty of solutions, ones we've known for decades. They aren't overly complicated. In fact, they're overly simple. First and foremost, we need to consume less and consume differently.

For the last twenty-five years, for example, I've been a vegetarian.

Initially, it was more a question of convenience than anything else. When you marry a vegetarian, you've got to be motivated to prepare your own meat for meals. And I wasn't. I was just too lazy. I'd eat meat when my wife, Rachel, and I went out, though. Like this insanely delicious steak salad at a Vietnamese place we used to regularly frequent. Cubes of steak marinated in red wine, salt, and garlic, placed hot atop a bed of cool, crisp greens. It's been twenty years, and I still have Pavlovian slobber in my mouth at the thought of it.

But the more I thought about it, and the more I thought theologically about it, the less I was able to sustain it.

From the standpoint of our God-given stewardship over creation, I couldn't justify it. We are "given dominion," sure. But the purpose of that dominion was to exercise care over the Eden into which God placed us. Eating the flesh of other creatures was not a part of that plan or part of what God called good (Genesis 1:29). If in Christ I am a new creation, and if Christ's work in me is to restore the breach established by our fall from Eden's grace, then not eating meat can be one way of personally affirming the healing of that rift.

Further, I feel that it is my responsibility as a Christian to minimize the amount of hurt and suffering I cause in the world. That's what it means to live according to God's law of love. Though chickens, pigs, cows, and the occasional possum are not as sentient or aware of their mortality as we are, they suffer nonetheless. They know pain, they know fear, and they die just as we do (Ecclesiastes 3:18–19). I personally prefer not to harm another creature if I don't have to.

And just as significantly in the context of climate change, a plant-based diet uses vastly less energy than a meat-based diet. It uses less land, requires less fossil fuel, and produces significantly less greenhouse gas on its own. Soy isn't known for its methane flatulence.

So I eat differently.

I also garden, which means for at least part of the year I'm getting a small but growing portion of food right there from my front yard. I compost organic waste, meaning that the clippings from grass and the leaves that fall from the trees on my property mingle with kitchen scraps. All of it becomes the dirt with which I garden. It's right there where I can walk to it. I can see the tomatoes on the vine outside my kitchen window as I write this.

Our home is a humble but sturdy suburban rambler, made back in the early 1960s from brick and wood and cinder block. It's 1,300 square feet, about half the size of the average new American home. We've updated insulation and replaced all the windows, so we're heating and cooling 50 percent less space. It's not at all fancy, and when it was the four of us, it was snug. But we've never really needed any more.

The appliances in our modest home are all selected for efficiency, and we use hyperefficient lighting.

Our primary car is a hybrid, a trusty Honda, and it gets more than fifty miles to the gallon in and around town. One tank of gas takes us around six hundred miles.

My secondary vehicle, the one I use to commute to church? A motor-cycle, which also gets more than fifty miles to the gallon, with the side benefit that it's both fast and fun to toodle about on when the weather turns lovely. It could do better on the particulate emissions front, but it still produces less than half of the carbon that burps out of the pipes of the average car.

None of this is hard. None of it requires any meaningful sacrifice. It costs me less. Our efficient life is comfortable, easy, and less expensive than an inefficient life. And it means, from a climate change perspective, that we produce slightly less than half of the emissions of the average American.

There are other ways we can shift the economy. We can continue the transition to relatively emissions-free renewables for our energy. That can happen at the industry level and locally as well. Installing solar panels on a rooftop, for example, costs no more than a modest used car, with the benefit of making a household energy independent.

We can ramp up the use of telework, pulling ourselves out of commuter hell and working remotely at far less cost. We could invest in public transportation, which is orders of magnitude more efficient than sitting in traffic in our individual vehicles. The list goes on and on.

Most of the folks I know who are justly alarmed at the shift in the global climate are more than happy to share such lists of to-dos. Here's how we'll solve the problem, they say. We must act now, they say, because we love the earth. We need to get this done.

Climate change is a test for humanity, and it's an open-book test. All of the answers are right there in front of us. But that doesn't matter.

We're not going to get things done in time. We're going to put it off. We're going to do other things instead. And by the time we realize that we shouldn't have been dawdling, that there was a task in front of us that had to be accomplished? It'll be too late. We are going to be stuck with a very different planet.

I am not a gambling man. But I'd put money on it. I'd lay down cash, and I'd win. Why? Because the test was yesterday. It's already happened.

22   David Williams

The time to respond to this global crisis was 1992, when the Union of Concerned Scientists issued its clarion call for an immediate response. "Move away from fossil fuels!" cried 1,700 scientists. "Manage resources more effectively!" "We have to act now, before it's too late!"

We didn't respond. We didn't show up. We didn't call the teacher and beg for an extension. We didn't check our report card or wonder why we didn't get credit for that class. We just continued on as if nothing had happened.

Now? It's too late. There's nothing that we can do to stop it because we can't change the past. We can lie about it to ourselves, which we're good at. We can tell ourselves that we have more time, that what we have put in motion can be stopped. But our wishes and lies are a fool's comfort. We can't undo what has been done. We have consigned ourselves to a harsher, more violent world, one that will test our mettle as a species.

Why? Why would humanity have done such a strange, nihilistic, self-destructive thing? There are numerous reasons.

Because we're small. Because we're afraid. Because we're proud. Because we're greedy. Because just as an object in motion tends to stay in motion, a culture's inertia is hard to change. These are all just part of what Christians used to call "sin" before we became anxious that it might drive away the seekers and seem all judgey. We failed to respond in time because we are mortal and sinful creatures. *Sin*: such a quaint term, so fusty and outdated and painfully accurate.

Ironically, we have done this to ourselves because human beings who call themselves Christians have made it possible.

This is our fault, which is tough. It is yours, and it is mine. We don't like to hear that. I certainly don't. But it is true nonetheless. And we need to come to terms with that before we repent and set ourselves toward the challenges of this difficult future. How, exactly, did we manage to do this to ourselves? What particular failings caused us to botch this so badly?

Let's look at that together for a few chapters.

PART II

WHY WE LET THIS HAPPEN

CHAPTER 4

SMALLNESS AND MYOPIA

Human beings are bright, wonderful, creative creatures. We care for our loved ones. We respond to our surroundings with insight and thoughtfulness. We're natural problem solvers, capable of applying our intelligence to whatever challenges we face. We are fearfully and wonderfully made, or so that Scripture goes (Psalm 139:14).

But we are also very, very small. The scope and scale of our living are shaped by our immediate context because of course it is.

I am no different. Take my typical Monday, for example. My Mondays are my guaranteed day off, the Sabbath day when I commit to doing no pastorly things, and they usually look much the same. I wake, and after setting my coffeemaker about the work of making my coffee, I walk the dog. I come back, drink a warm cup of go-fluid, and read the paper. I check media and write for a while. I water the garden and perhaps do a little bit of weeding. Then I look to the debris field that the house has become over the weekend as Rachel and I have trundled about doing things.

So I settle in, and I gently sort the chaos into a semblance of order. The pile of dishes in the sink slowly disappears. Counters and tabletops are cleaned, and bits of paper that do not belong where they have been placed are either recycled or moved to a more logical location.

It's not particularly impressive or unusual, to be honest. I am not building a reputation. I am not wowing the world with the wonder that is me. I am not shaking the foundations of injustice with the power of my prophetic witness.

I'm sweeping the kitchen floor.

Again, to be utterly honest, I like sweeping the kitchen floor. There, on the floor, lies the detritus of several days in the life of our little home. A bit of cereal here, by a clod of dirt from the garden. A niblet of dry dog food there. I take the broom, and I sweep those leavings into a little pile, which is then neatly ingested by our Dustbuster. It is a satisfying thing.

Then I mop. As I mop, the stains of the days come up off of the tiles. It is pleasurable, simple, direct, and with a clear result. The floor was dirty. Now, for a while, it is clean. It has returned to order.

I am like a dream moving through the mind of my sleeping home, sorting its thoughts and memories into order again. It is not a particularly impressive thing, this cleaned kitchen of mine. But we are not impressive creatures. *What am I doing?* I think, sometimes, as I clean. *Make a name for yourself! Pour your energies into seeking acclaim!*

But why? Why would I tear myself to pieces chasing after something that is no more meaningful than the thing I am accomplishing?

I look to the heavens, to the great deep of creation, and I see how little personal glory would matter. If I were the emperor of the world, with a vast robot army and undisputed command over all of humankind, and all loved me and despaired, I'd still be a speck on a speck in the vastness. I consider the complexities of the subatomic realm, in which our every movement is irrelevant to the elements and energies that compose us. What is fame, on that scale? What, even, am I? Less than a breath. A cleaned floor is, on either scale, no more and no less significant than the entire life of Alexander the Great.

So I look to the scale on which I live and take pleasure in my kitchen, cleaned and mopped and straightened. What an excellent thing, I think.

And it can be. Most of our human life needs to be lived like that, in a way that is humble and mindful of what is most immediate. We tend to our children, wiping the snot from their noses and changing their diapers and delighting in their growing souls. We go about our labors of the day. We talk with a friend or run to the store. These things exist on our scale, and we can easily perceive them and respond to them. They demand our attention, and we can see them, and they are obvious.

Climate change is not such a thing. It doesn't exist on the scale of the immediate. It doesn't feel present, not in the same way that the rest of our life seems present. It does not happen on a human scale temporally or spatially. It is both too large and too slow to spring to our attention.

Our memories only go back a short distance, and slight changes year to year just don't register. On a historic scale, on the scale of our world and our ecology, the changing climate has been an explosion, a catastrophe that is playing out across ten generations of brief, fleeting human life.

We personally don't notice that, most of the time. Things are just a little different, just slightly changed. A few more trees die from heat stress. The winter is a little warmer. Oh, that's weird, we say, but then we go about our busyness. We can't see something moving at such an incremental pace.

It's like watching a child grow. If you see them every single day, it's easy to miss how quickly they're blossoming into adulthood. But when you go to finally visit your brother and sister-in-law who live far away, that towering fifteen-year-old nephew is suddenly a completely different person than he was three years ago.

Our world's pace of change is terrifyingly fast by biological and ecological standards, but human life is just a flicker against that scale. We just don't see how different it is. For that, we need our imaginations.

Go back three life spans, three generations of human beings, to the life of my great-grandmother. The year, 1919. On the mantle over my fireplace, there is a tintype of her as a young woman, her eyes soft and inquisitive. She lived a hundred years back, just a blink of an eye historically.

Now take my great-grandmother and place her deep into a forest near my house. There are no homes visible from that forest, which sits in the middle of a large county park. It's deep and quiet, trees upon trees, with only a few houses nearby and out of sight. We can imagine that the view from that forest is much the same as it would have been from any forest in the last century.

But she wouldn't for a moment doubt that everything had changed. If she stood in that forest in the still of the morning, she'd hear that it was different because the air is never quiet. There's a roar, the great, distant, ever-pealing thunder of hundreds of thousands of cars on the move. Millions of tons of metal rolling over rock aren't quiet, even if they're riding on nice new tires. For those of us who live in the sprawl of the 'burbs, it's just background noise. We tune it out. For those who lived a hundred years ago, it'd be as unsettling as the sound of a massive lava flow nearby, or the roar of an avalanche.

If you set my great-grandmother into that forest at night, and she looked up, she would see a sky that was terrifyingly unfamiliar, one in which the heavens had been almost entirely erased. The stars themselves would seem to her to have been blotted out by a great omnipresent glow, as if all around, the world was on fire. We don't notice that because it has just always been that way for us.

Set into those acres of woodland is a clearing, a sweet, wide field on a gently sloping hillside. It's just about the best sledding hill in the area, or it was back when we got regular snowfall. If you set my great-grandmother into that field in the late fall, she'd see a few straggling geese now and then, flying in a small vee across the sky. The great flocks of migratory birds that were part of her world are all gone. In the spring, she'd see only a handful of butterflies, and the flowers would be empty of bees.

We don't notice these things because they fade away year by year by year, and we're too small and too busy about our little lives. Only very rarely do things happen that startle us, that take us by surprise.

Back in 2012, I was taken by surprise by just such a weather event. My family and I spent a few days without power, sweltering through heat made tolerable by a decent emergency plan in our household. Because we live in the DC area after 9/11, the anthrax scare, and the DC snipers, we're prepared for most emergencies. We had water, and a large reserve of canned food we didn't even need to touch, and a propane-powered grill that's provided us with hot meals and hot water for French-press coffee. Oh, and a handy little emergency battery power supply.

What was unsettling about the event was that for all of my then forty-three years of life, I did not have a frame of reference that let me grasp what was happening. It came on a Friday night when we were expecting storm activity. But when things happened, it didn't meet any of my expectations. We woke, startled, and with the waking, it was clear it was not an ordinary storm. The house thrummed and shook as wind and rain and hail battered it. The trees in our neighborhood rocked and shook wildly, and the sky was lit with endless, constant lightning and thunder.

As the power went out, it felt for a moment like we had inadvertently wandered into some apocalyptic film or a scene in some old *Star Trek* episode where a high-energy plasma storm was sweeping across the Enterprise: "We've lost all power, Cap'n! She's dead in the water!"

Then there was the sound. The thunder was so frequent it blended into a roar, which mingled with the overwhelming shrieking of the wind, which mingled with the rice-crispies-of-the-Titans sound of trees snapping, crackling, and popping. It was genuinely alarming.

"What is it, Dad?" shouted my panicked younger son as we hustled downstairs to the security of basement cinder block with the dog and the emergency power brick in tow. "I don't know," I said, because I didn't. It wasn't a tornado. The wind wasn't right for that. The thunder wasn't right for that. It wasn't a regular storm. Though I follow weather avidly and read about interesting weather phenomena, this was a baffler. I was at a loss. I genuinely had no idea what was happening. "I don't know what's happening" is a hard thing for a dad to say to his frightened child in a crisis.

For another few minutes, the whatever-it-was raged, diminished, then raged again . . . and then passed on.

Now I know the name of it. Millions of Americans got to learn a new word: *derecho*. A derecho is a heat-driven atmospheric energy event. To visualize it, I think of something that's basically a tornado laid on its side, stretched out into a bow over one hundred miles or more, that just blasts and tumbles its way across the landscape—in the case of this one, a long way across the landscape. It's not precise. It's actually kind of wrong. But that's the net effect.

I rode from my home in the Virginia suburbs to my church in rural Maryland two days later. Picking my way through thirty-five miles of nonstop devastation, I found myself thinking many things. One was that my family is going to need to upgrade our emergency plan. And another was that my frame of reference was going to need to change to take account of this new, strange, violent world.

When I arrived at church on Sunday morning, I had only the vaguest idea what the day would look like. There was stuff on the schedule, of course. We had a youth mission team headed off to Kentucky that was going to be blessed. It was a Communion Sunday. There was a postworship Bible study on the docket.

But with the entire region suddenly shut down, none of that was certain. Communications were critically compromised. There was no power, and we were in the middle of a heat wave. My phone was a pretty but almost useless brick. In the few moments the overburdened cellular network had let me in, I'd managed a few emails and texts with our stalwart worship and arts elder, in which we agreed to keep it simple and play it by ear. Maybe just a prayer, some lemonade, and a send-off for the mission team. We'd see.

I arrived early, wending my way through debris and stopping at one point to clear a country road of a fifteen-foot branch that was blocking most of it. Things at church were better than I'd expected. The huge, graceful tree sheltering the fellowship hall was undamaged. The not-quite-as-huge tree next to the manse was also intact. There was no power, of course, which meant that our small sanctuary could have passed for a brick Easy-Bake Oven.

More and more souls arrived, as did the communion elements, followed by more souls. Though I was ready to do just a prayer, the Spirit felt there for a shortened service. Chairs formed an impromptu semicircle in the fellowship hall, which was at least ten degrees cooler than the sanctuary thanks to the sheltering wings of that old tree. A bustle of folks snagged stacks of hymnals. A folding table received the elements. A lectern slid into place.

And so my wee kirk worshipped, blue bulletins fanning all aflutter. The hymns I'd selected were—thank the Maker—all well known and easily sung, and though we were a cappella, the room filled with strong voices and harmonies. It was hot, but the room was filled with good cheer.

As I preached from the handwritten text I'd re-created by candlelight the night before, the sweet-salty taste of sweat touched my lips. I lacked the necessary kerchief, so instead of mopping my brow with a South Georgia country pastor flourish at opportune pauses, I just let it run down my face. The flavor of a good honest sweat was still there as we shared the Lord's Supper, mingling favorably with the sweetness of the juice and the slight tang of the sourdough.

After the service, the cool of shade-giving trees, watermelon, and lemonade made the lingering a pleasure, one that was a welcome respite from our hot, dark homes and yards filled with debris that would have to be moved. In the conversations afterward, I asked my congregants, Have you ever experienced anything like that? And although they were drawn from all around America, and many were on the older side of things, none of us had.

It's not that derechos never happen. It's just that they are rare, markers of an unusually high level of energy in the atmosphere, the kind of event you can easily live a whole human life without ever encountering. I don't think we'll have that luxury on our new, more violent planet.

Against the intimate scale of our lives, it is easy to miss those things and to misunderstand them. We recover and then go back to bustling around, consumed by our busyness and activity, little ants diligently bringing the grain back to the nest, bees laden with pollen as they fly to the hive.

Which is why, when the warning bells of the climate crisis first sounded, we didn't hear them. We were just too small, and it was too big to see.

So we didn't act, and now it's too late to avoid it.

CHAPTER 5

POWER

None of us likes to feel helpless. We want to be competent. We want to be capable. And we enjoy power.

Power—the ability to influence and move the world around us—is one of the great desires of humans. Being a human myself, I understand this viscerally. I like power. I like that feeling of competence and potency that comes with being able to shape what I encounter.

For my birthday a couple of years back, my dear wife gave me the present of power. She knows that my Scots-Irish blood means I am a creature of peculiar contradictions. I am, being a Scots Presbyterian, a lover of thrift, frugality, and the practical. My inner Irishman, however, enjoys a good raging hooley, a bit o' the wildness that makes life's story one worth telling.

I've learned, over time, that it's best to let John Knox make most of the significant purchasing decisions. So although an efficient hybrid and a practical used minivan do their stalwart duty in our driveway, my heart longs for more exuberant things. I pine after fast cars, just as I have since I was a little boy.

"Your present: go rent a fun car again," she told me on my birthday. She's done it before—the gift of a fun car rental for a week—and it's a favorite gift.

As it so happened, inbound was a great beast of a snowstorm, the forecast peppered with the increasingly familiar adjective: *historic*. The tolling bell of weather panic was ringing from every virtual steeple. As other

Washingtonians scurried about like tiny Tokyo denizens before a giant, rubber-suited man, I contemplated the most optimal nonmilitary surplus vehicle for dealing with the snow. *Surely this is a fine time for getting a big ol' four-wheel-drive truck*, whispered me inner leprechaun. Maybe something red. Big, red trucks are nice, they are.

But my inner Scotsman was stubborn, as Scotsmen tend to be. It was still so frivolous, so pointless, such an expense. How could I justify it? Why not just settle in and deal with the storm as I'd dealt with every other storm and just sit it out?

Then came the call from my parents. My dad, whose hip had recently been replaced, was in the hospital. The new hip had popped out, and after several painful hours in the emergency room, they'd jammed it back in. He was incapacitated and back at home with my mom. With that "historic" blizzard rolling in. I needed to be able to get to them. I could not be helpless. Period. My Scots and Irish selves, with their fierce love of kin, agreed.

So on the day of the storm's arrival, I pulled into our driveway with a rented Big Red American Truck—a Ford F-150 XLT SuperCrew 4×4, to be precise. That first morning, as the blizzard howled, I cleaned the great red beast off and prepared to test it out on the eighteen to twenty inches of unsullied powder that covered our suburban road. The snow, up just above my knee, seemed impossible, impenetrable, so deep it was a challenge to walk.

I've always been a confident snow driver. Heck, I even enjoy it.

But looking at the depth of Snowzilla's first evening's leavings, I was convinced that I'd made a mistake. I'd lived nearly half a century in the DC area and never seen a snowfall this intense. The truck was going to be a colossal boondoggle. I'd pull out, and get stuck, and that would be that. It would be lovely driveway candy for as many days as it took them to plow us out.

I was wrong. In four high, with traction control engaged, and with 1,500 pounds of snow packed down in the bed right over the rear axle, the truck was unstoppable. That big Ford just . . . pushed. And the snow got out of the way. I made two runs through the neighborhood, carving nice deep walking tracks into the snow, perfect for walking our dog.

As the blizzard roared on, I got out every hour or two and kept tracks running through the neighborhood. It was, occasionally, a bit technical. And more than a little fun. I got through, pounding through four-foot snow berms. The truck and I could not be stopped. Six thousand five hundred pounds of 4×4 with mass quantities of torque just gets where it needs to get.

So I ran errands, checked in on my parents, and brought groceries to some elderly family friends who were stuck in their home. We went for dinner at my in-laws as the storm roared, shuttled my teens to help with shoveling afterward, and surveyed the scope of Snowzilla's impact.

It felt great. I felt powerful. Unstoppable.

A day passed. Then another. The roads got plowed. Not all of them, but most. I ran errands. We visited my parents and my in-laws. I stopped to help the stranded, including a plow that got stuck on a hill I'd traversed just three minutes before.

It was utterly enjoyable.

But on the last day I'd rented the truck, as streets reached the point where I could easily traverse them with one of our more frugal vehicles, I began feeling, well, odd. Here I was, in a glorious beast of a truck, and . . . I didn't need to be. I'm not a farmer or a contractor. I don't live in the Upper Peninsula or Fargo. It felt like excess capacity, like the kind of meal that hits the spot after a day of hard labor or physical activity but that, if you've been sitting around all day, just leaves you feeling off. And as fun as it had been, I found myself eager to be out of it and back into something that more reflected my actual needs. Something that got nearly fifty to the gallon instead of barely fifteen.

Power, after all, can be a dangerous thing if we become too used to it.

And we have gotten used to it. The oil crises I faintly remember from my childhood in the 1970s are long forgotten, and cars have gotten big again. Heck, we don't even drive cars anymore. We don't. We drive SUVs and trucks in the United States. In my neighborhood, over the course of the last twenty years, I've witnessed that change. Driveways that were once home to cars are now filled with four- and all-wheel-drive SUVs. I have neighbors, perfectly lovely souls, who commute to their offices alone in pickups that are larger and less efficient than the one I rented.

We like them because they're big, and they're spacious, and they're powerful. They appeal to our desire to do what we want, whenever we want. That they are significantly less efficient means almost nothing against that desire to be able to go and do, unfettered by the world's restrictions. We have become acclimatized to that power. We have integrated it into our everyday lives. Power is the manner in which we are accustomed to living, and that's difficult to shake.

Here's another example. It begins with the most preposterously over-powered vehicle sold in the United States today: the Dodge Challenger SRT Hellcat. It is the Platonic form of the muscle car, a huge slab of overpowered absurdity, arguably the high-water mark of the guzzoline age. It only has two doors but seats four comfortably, being a huge hulking beast of a car. It has a supercharged 6.2-liter V8, which cranks out more than seven hundred horsepower, more than six hundred foot-pounds of torque. If from a dead stop you stomp on the accelerator, hard, the Hellcat will just spin the back tires until they explode. There was a recent test by a major automotive magazine perversely designed to see which car could burn through a whole tank of gas in the least efficient and most eco-harmful way. In that test, a Hellcat was put through one full-throttle quarter mile after another, over and over again, until it ran out of fuel. It managed to get four miles to the gallon. Four.

This car is the absolute antithesis of creation care—blaringly, ragingly, willfully so. And Lord help me, but the fourteen-year-old boy in me can't not kind of want it.

Every two years, my earnestly progressive denomination gathers and earnestly discusses Issues of Great Concern. That includes, typically, climate change and our use of fossil fuels. Presbyterians from all around the country gather to share and talk and dream about what it means to follow Jesus. That's not a bad thing. At the last several meetings, there's been talk of divesting from fossil fuels and of all manner of ecological initiatives.

But there is an irony there, one that is inescapable, frustratingly so. To get to this meeting, most folks fly. To fly, you must use considerable amounts of fossil fuel to accelerate a large mass to nearly five hundred miles an hour. That fuel is burned high in the atmosphere.

To be fair, commercial airlines are much, much more efficient than they used to be, as aircraft now use half the fuel they did in the 1970s. If you fly on a commercial flight now, your carbon footprint and energy consumption are about equivalent to driving a ten-year-old Honda Civic. Meaning: that time in coach is as environmentally friendly as you driving a vehicle that gets thirty-five miles to the gallon on the highway. Not bad.

That's particularly not bad against the average vehicle in the United States, which gets twenty-one miles to the gallon on average on the highway cycle. The Hellcat, when loping along the open road at eighty miles per hour in the top gear of its eight-speed TorqueFlite transmission? It puts down about twenty miles per gallon—pretty wretched for a two-door, and almost 57 percent worse than flying.

Here's the thing: that entire metric changes when you aren't driving alone. Let's say I were to have taken two of my fellow presbyters on a three-day road trip across America in a Hellcat. It's a big, spacious 'Murikan car, after all. Our net energy consumption and emissions per person would have been the equivalent of a single individual driving a vehicle getting sixty miles to the gallon. Meaning a transcontinental journey in the most absurdly overpowered production muscle car in the history of internal combustion engines would be a nontrivial 70 percent more eco-friendly than flying.

Most attendees of this national event have no choice. You fly because this is a large nation and crossing it over land takes days and days. You fly because you are too busy not to fly. That's how we live. Busy, busy, busy bees. And to manage that busyness, to accomplish all we need to accomplish, we need power: the power we get from the rich, dark energy density of fossil fuel.

But there is an inherent ethical dissonance between the jet-age convenience of a fossil-fueled air journey and the argument that we need to stop using and investing in fossil fuels. A coherent ethic needs to encompass both our systemic and individual choices.

One can, of course, still make the argument for divestment having gone and done flown yerself there. You can still make the argument for environmental stewardship while recognizing the dissonance between the society we inhabit and the one we wish to inhabit. It's not hypocrisy if you're aware of it and working on it. It's not hypocrisy if we realize that we aren't perfect and are willing to name our imperfections.

But that love of power, that need to be endlessly dynamic and overcoming and doing as we wish, whenever we wish, however we wish? Our hunger to dominate and our appetite to rule over both creation and one another is far harder to change. The hunger for power over others is the heart of human sinfulness, after all.

So we roar onward, feeling our power, feeling proud, and we somehow manage to forget about the world that is hardening itself against us.

CHAPTER 6

INERTIA

Back when I was a small boy, summers meant that we were traveling to visit family. Around the fourth of July, we'd travel down to Athens, Georgia, for a week to visit my maternal grandparents. We'd spend a sleepy week going to the pool, watching fireworks, and chasing fireflies. We'd also tag along with my grandparents as they went to visit friends.

One house in particular stuck in the mind of little-boy me. Before he retired, my grandfather had been head of the math department and the dean of the graduate school of the University of Georgia, and one of his mathematician friends lived on a large, multiacre property a short drive outside of town. Visiting the Balls was heaven as a kid because (1) Mr. Ball had a very large Kawasaki motorcycle that I would be permitted to climb on; (2) the Balls had their very own stocked pond, where we would go fishing for the bass we'd fry up for lunch; and (3) as Joe Ball was a mathematician, the Balls had a home computer, on which there was a game. A computer game.

Understand: this was the mid-1970s, a couple of years before the Atari 2600 went on sale. Just having a computer was a rarity. A *game* on a computer? That was pure magic, an amazing thing in the summer of 1976. With a little instruction and cautions about what I could and could not touch, I was permitted to play it. It felt, honestly, like I was a tiny, hairy-footed hobbit child and Gandalf had just handed me one of his fireworks.

The game: *Moon Lander*. *Moon Lander* was the precursor to the classic game *Lunar Lander*, and it's wildly simple by today's standards. Against a simple green vector background, you control the thrust of a spacecraft as you attempt to land it on one of several pads on mountainous terrain. You have to manage fuel consumption and momentum, carefully adjusting your altitude with a delicate touch if you want your little lander to survive.

Once the simple but unforgiving physics of that game had you, it was very difficult to avoid the simulated Newtonian consequences. That little green glowing craft liked to stay in motion, and if you let it get away from you, you'd reach a point where even full thrust couldn't prevent it from plowing into the sketched lines of the world below.

I was in third grade, and so I did a whole bunch of crashing. Sometimes on purpose. Often on purpose. I'd apply all of my thrust at once, and early, so that the little ship soared upward and then came plummeting down. Firing the engine helplessly, with the fuel running low, I'd make tiny panicked astronaut noises.

An object in motion, after all, likes to stay in motion—which, when you finally grow up and aren't just playing around in the virtual world, can be a little less fun. Newton's first law should be familiar to all of us from science classes back when we were kids. It's the law of inertia: the idea that in order to change the state of a thing, you need to apply energy. Once an object is moving in a particular direction, you need to apply force—or intention—to get it to do something different. Similarly, if an object is just sitting there like a lump, it's not going to do a dang thing unless you apply force. Sir Isaac intended this definition of inertia to describe the world of inanimate objects, but it has a peculiar resonance in our personal lives and in our societies. Change does not just magically happen. We do not wake up in the morning fifteen pounds lighter and more toned. You don't get the Apollo program without a significant commitment of focus and effort. You do not stir a nation to overturn Jim Crow without focus, energy, and commitment. Cultures and peoples only change if they overcome their inertia. And that is harder for some societies than it is for others.

Turning around an entire economy is considerably harder than changing the direction of a tiny little vector-graphics spacecraft. Our reliance on fossil fuels for energy is nearly complete. We need them to move goods, to heat our homes, and to produce the cheap and easy industrial food that lines our expanding midsections. We require fossil fuels to travel to our jobs, where the machines that we use to work rely on energy produced, more often than not, by fossil fuels.

Making the turn to renewable, carbon-free sources of energy requires time, foresight, effort, and time. Particularly time. It's entirely possible, if we were committed to putting in the effort—because overcoming inertia requires effort. It requires energy, commitment, and planning to change direction.

Could we have done it? Sure. Sure we could. Other nations have made that choice. Let's look at one: Norway.

I recently traveled to Norway with Rachel because, well, Norway is beautiful. It's also really, really chill—one of the easiest-going countries I've ever had the pleasure of visiting. The Norwegians don't stress about things in the way that we Americans do. Part of that, I'm sure, is that they don't worry that they'll go bankrupt if they get sick. A large proportion is just a part of their national character. They're not anxious or controlling people, in ways that sometimes boggle the American mind.

On our journey through Norway, my wife and I drove down stunning, perfectly maintained country roads in our tiny rental Audi on our way to see one of the many must-see things in Norway: the Kjeragbolten.

To get to the Kjeragbolten, you park your vehicle and hike. It's not a short hike. Neither is it easy. It's surprisingly hard, in fact. You're traversing, in many places, rock faces at a nearly forty-five-degree incline, pulling yourself up by a chain affixed to the face. The tumble down that incline would be fifty to a hundred feet, and it's all rock, but hey—no problem. Families are doing it. Kids are bopping along on their own. As far as Norwegians are concerned, if you can't do it, then don't. And if you're not into heights, the Kjeragbolten is probably the wrong hike for you.

The Kjeragbolten itself is a rock. It's a big rock, about the size of a cube van. It's jammed in between two cliff faces, forming a small bridge. That small bridge sits over a drop that goes down—straight down—just over a kilometer.

There are no handrails around the Kjeragbolten. No signs. No warnings. No liability waivers. Just a long line of Norwegians and European tourists, kids goofing off, chatting genially with one another as they wait next to a thousand-foot precipice for their turn to clamber out onto the rock to have their picture taken. One after another they do, smiling and waving above a yawning, hungry chasm, an inescapably lethal drop all around.

Norwegians don't worry about it. Again, if you want to do it, do it. If you don't like heights, don't do it. And don't be a moron, or even our Scandinavian health care isn't going to do your splattered remains much good. It's pretty simple and utterly devoid of fearfulness, in a we-used-to-be-Vikings-dude-what's-the-big-deal kind of way.

Norway's a peculiar place, and an even more deeply peculiar thing from our American perspective is their approach to energy. Norway is, after all, an OPEC (Organization of the Petroleum Exporting Countries) nation. They've benefited from the vast reserves of oil under the North Sea, a reserve that has poured billions of krone into the Norwegian economy.

On the West Coast of Norway lies the city of Stavanger, which is both at the heart of the Norwegian oil industry and a remarkably charming town. In Stavanger, the wife and I spent a morning at the Norwegian Petroleum Museum. It gives fascinating insights into the role of fossil fuels in Norwegian culture and the herculean engineering efforts that make drilling in the North Sea possible.

Two things were particularly striking about that museum, both highlighted in an exhibit discussing the future of petroleum in Norway. First, up on the wall was a large LED display showing two numbers. Visualize, if you will, one of those "deficit counters" that American fiscal conservatives use to point out our catastrophic national debt. Here's how much we owe as a nation, and here's how it breaks down per person. We look at those numbers, groan, and go right on electing profligates to office.

Because all fossil fuels under Norwegian soil are viewed as belonging to the nation, Norway has the opposite thing: a national surplus counter, showing the hundreds of billions of dollars they are in the black as a country, which is broken down to show the share of every Norwegian citizen. It's more than a million dollars per person, which is a little painful for an American to see.

Second, the Norwegian Petroleum Museum is brutally frank about climate change. If you had to summarize that part of the museum, you'd say, "It's real, it's caused by human activity, and we're contributing to it. As a nation with significant oceanfront, we're going to be impacted by it, and so we need to prepare." There's no prevaricating, no ideological groupspeak, no profit-hungry self-interest.

Faced with the impacts of climate change, Norway is preparing. They've got a national plan for ramping down oil production. Though they're an OPEC nation, they tax their petroleum intensely, as I discovered when I went to fuel up that little Audi. Doing the math—between krone per liter and dollars per gallon—on the fly at that gas station was a little traumatic.

The Norwegians are using their surplus to ensure that they have a clean and efficient system of public transportation. In cities like Oslo and Bergen, I saw more Teslas than I've seen anywhere outside of a Tesla dealership, part of the aggressive move away from internal combustion.

Norway now gets more than 99 percent of its energy from renewable sources. They're as prepared as a nation can be and have turned their wealth as a nation toward that end. They've been preparing for decades because that kind of transition takes time, even in a small and wealthy country.

But in the United States, we haven't. We're a whole lot larger. The gross domestic product of the United States is around thirty times the size of Norway, and our sprawling mix of regional cultures makes it harder for us to find the consensus we need to change directions. I mean, we *should* be more flexible, right? Being that we're not "socialist" and all. A land that talks so endlessly about liberty should be free to do things a new way, right?

But we aren't, because changing the patterns of our lives doesn't come easily. We are creatures of habit, and we like the familiarity of day-to-day sameness. We succumb to inertia, and we aren't willing to put in the effort that will be required to change direction. It's not that we couldn't. Like all changes in life, it's a question of will, energy, and effort.

We don't want to put in the effort to change. We're not interested in trying because trying might be hard, and it's confusing, and we're hearing so many different things, and it all seems like too much.

So we don't bother, and we roar along with the flow of things. And with the cries of tiny astronauts, that impact waits in our future.

CHAPTER 7

GROWTH AND GREED

O ur entire culture is founded on a peculiar idea: growth is a constant that will continue forever. Nothing is static, and everything is changing and expanding and increasing. It is our assumption about how things are. It is our assumption about how things will be.

For most Americans, this is understandable. We've watched growth roaring around us—everything getting larger, woodland and prairie becoming fields and farms, fields and farms turning into strip malls and subdivisions in what feels like an endless process.

In my own suburban community, where I've lived for more than forty years of my life, nothing feels quite the same. The drive-in theater that sat by the side of a nearby road in my childhood was leveled and replaced with a multiplex. That multiplex was leveled and replaced with what I like to call an insta-city: a dense, mixed-use development, with theaters and shops and restaurants and town houses starting in the "affordable" low nine hundreds. That's not nine hundred a month, my friends. Those town houses are $900,000.

The county where I live has added nearly five hundred thousand souls since 1975, as decades of debt-fueled government spending have poured money into the DC Metro local economy. Growth is what I know. Growth, we are told, is progress.

But is it?

I look out across the endless expanses of tarmac and idling steel and flesh stretching to the horizon, and I think, *This isn't progress.* Progress should feel . . . graceful. Progress should be . . . moving. As the suburbs have grown around the easy availability of petroleum, cities and towns and communities built on a human scale have vanished. The suburbs have flourished and spread like a fungus on the inside of an unwashed thigh.

Oil has meant growth, without question. It has driven massive expansions in transportation infrastructure and underlies all modern commerce and industrial agriculture. It is the engine that made the explosive human population growth of the twentieth century possible.

But I wonder: Is it good growth? Is it healthy growth? Or is it the growth that kills? Because some growth kills.

Agent Orange is my favorite ecological noncancer metaphor for growth unto death. That narsty substance, in the event that you don't know your twentieth-century history, was a herbicide sprayed onto the forests of Vietnam by fleets of American planes. To kill Viet Cong, or so we thought, we had to deprive them of cover . . . so we killed their jungles. Agent Orange was the plant toxin we used, and it works in a very interesting way.

It simulates a plant growth hormone and essentially causes most broadleaf plants to go into a period of explosive and unsustainable fecundity. After spraying, leaves would grow huge. Fruit would be immense, distended, mealy, and inedible. A jungle poisoned by Agent Orange would, for a short while, be an alien wonderland of insane, outrageous production. And then, having exhausted itself and burned through all of its natural reserves, the jungle would die.

For one hundred years, we've consumed fossil fuels, and from that easy but finite energy we have grown explosively. But the age of oil is limited and comes with unanticipated cost to our world. As I look out across what human society has become, I do now and again wonder.

Are fossil fuels our Agent Orange? Is the growth they have caused the kind of growth that burns out and destroys? Because at the far end of this era, they will no longer exist. You can't replenish them, not unless you wait a million years or so. The easily portable, easily accessible ~~energy~~ density of those fuels will be gone forever.

Were we a more rational people, or if we lived with the foresight that rises from wisdom, we'd be taking this seriously. Aware of both the damage and the finite supply of fossil fuels, we'd be cutting back. We'd be managing our transition.

But we are not, because there is money to be made *right now*. What we are hearing instead is that it's a time of energy abundance. So crow the pitchmen for the oil and gas industry, eager to spin the current decade of fracking-driven oil glut into something wondrous. A new age for America, part of an exciting time of expansion and growth, driven by ample new supplies of energy. Cheap gas! Big cars! Heck, don't even drive a car—drive your very own luxury truck! It's 1969 again, baby! Abundance!

And on the one hand, taken as a snapshot of this moment in time, that is true. But it is a strange truth because it masks a larger reality.

That reality is that we have arrived at the shores of a new world. It is a world both fiercer of climate and soon to be devoid of fossil fuels. Oh, we have them *now*. But within what I can reasonably expect to be my children's lifetime, our transportation system—and our agriculture—will have to be completely new.

This would have been doable if we'd put in the effort back when the scale of this crisis became obvious. But we did not. The global storehouse of fossil fuel energy is just as finite as it ever was. The only change is that we're pumping it faster, aggressively flushing the last drops from the grudging shale. Within the next fifty years, our entire energy economy will change, whether we want it to or not.

This is reminiscent of something from the beginnings of the American story. What this time on our new planet reminds me of is the experience of the first American colonists.

Where we are right now as a planetary economy is Jamestown, and it's the autumn of 1609. We are those colonists. We can't grow any more food because the growing season has passed. We haven't made friends of the Indigenous folk around us. What we have in the provisions that we've brought from England, and what remains from our sparse harvest, will have to get us through the long winter in a new land until we are able to figure out a way to be self-sufficient.

What the sheiks and the oligarchs, the executives and the lobbyists want us to believe is that our finite storehouse is justification for a time of feasting. "Look how much we've got in the storehouse," they smile, as they overproduce, cut prices, and tell us all to eat our fill. "Wow! So much! Time for a celebration! Yay us! Dig in! It's so cheap! Happy to sell it to ya!"

Which in Jamestown, in the fall of 1609, would have been sort of true—but mostly not. That colony barely, barely survived a time of hunger and privation. Others, like the Roanoke colony, simply vanished, dying on the hostile shores of a new world.

What we are being told now is wildly unwise, the deluded foolishness of profit-addled quartermasters. And profit they have, with the new boom in consumption driving fossil fuel industry profits to new heights in 2018, with tens of billions in income domestically. ExxonMobil alone, in 2018, had profits of more than $20 billion. Those profits drive shareholder value, which is the goal of any business oriented toward profit maximization. That, simply put, is the goal. There is no other.

Profit doesn't just provide shareholder return. It is, for many OPEC nations, the foundation of their economic power. In 2019, the world's most profitable company was Saudi Aramco, which brought in US$88.2 billion in net income. For context, that's nearly 40 percent more profit than Apple yielded.

Those yields support the Saudi monarchy, which uses them to maintain control of their populace. They aren't eager to step away from production until it becomes absolutely necessary. The Russian state-owned oil company Rosneft saw lesser profits but still brought the equivalent of nearly US$6 billion into the coffers of the Kremlin in 2018.

The idea that states and corporations will simply step away from production is absurd. There's simply too much potential wealth available, and human beings are human beings (unless, apparently, they are Norwegian).

It's the challenge of an economic system that is oriented toward short-term profit, in which the yields this year or this quarter are what matter and the "big picture" is meaningless. It's the challenge of a system in which growth and profit are fetishized and our greed and hunger for mammon are rewarded.

That has always been the human condition, and there's simply no reason to believe we'll change until the crisis becomes obvious. It's like that moment, blundering lost through the forest at twilight, when you crash through a bush and realize you've stepped off the edge of a hundred-foot precipice. Sure, you're now aware of the problem. But it's too late to do a thing about it.

So we have already blundered into an inescapable crisis. The reasons for our failure, as we've seen in the last few chapters, are just part of our flawed human nature. We are small and shortsighted. We love power, and we're prideful creatures. We have trouble changing our habits of life. And we're greedy and obsessed with more, more, more.

Our failure to respond to this crisis, in fact, could just as easily be defined and described as the fruit of the seven deadly sins. We are in this position now because of our lust and our gluttony, our greed and our sloth, our wrath and our envy and our pride.

We're sinful creatures, in other words, and as a species, we are in the process of harvesting the fruit of our sin.

What will that harvest look like? What, exactly, have we gotten ourselves into? Let's take a look at that now. What will this harsh new world be like?

PART III

HOW WE FACE THE CRISIS

CHAPTER 8

WHAT HAPPENS NEXT

We know where we are now. Odd weather events, storms that are "historic," floods that are "biblical." The ice covering the polar seas is melting. Record-breaking temperatures occur almost daily across the planet. There's precisely zero question that the earth is warming and that human activity is causing it. The climate crisis is right here, right now.

But what does that mean in ten years? What does that mean in twenty? What will it mean in a hundred?

We don't know. We just don't. Not with any precision, anyway, because the transition of our climatic systems and ecologies represents a complicated and chaotic event. There are worst-case scenario assumptions that involve a cascading cycle of failure, each one grimmer than the last.

In the frankly terrifying 2017 article in *New York Magazine* that provided the grist for a subsequent best-selling book, journalist David Wallace-Wells laid out the scenarios toward which we seem to be moving. The article had the totally-not-alarming title "The Uninhabitable Earth," which was definitely a little on the distressing side. In fact, it's the sort of article that would be nice to dismiss as manipulative fearmongering or some cynical effort to get people riled up and angry and tweeting about your article. But it wasn't. These aren't the panicked projections of some rando who tweets conspiracy theories from his basement. The article simply articulated the dry projections of climatologists and biologists, the trained scientists who actually study and understand our ecology and climate systems. Those possible outcomes were the farthest thing from pleasant.

They begin with the threat of uninhabitable equatorial regions, as skyrocketing temperatures and humidity make it impossible for human beings to live, move, and work outside during substantial portions of the year. Wallace-Wells cites a groundbreaking study produced by the National Academy of Sciences, which noted that within current projections, the wet-bulb temperatures we will begin to encounter are enough to cause death by hyperthermia in healthy human beings. With ferocious heat and much higher levels of ambient humidity, we can't cool our bodies. We—and our livestock—simply cook from within. It's not quite the surface of Venus we're talking about here, but there's the strong likelihood that large portions of the surface of our new planet will be rendered uninhabitable for substantial portions of the year. This includes large swaths of the American South.

The grim litany in Wallace-Wells's summary of likely eventualities goes on. There is every reason to believe that food supplies will be compromised, as agricultural yields collapse under heat stress. Our current reliance on grains for livestock feed just won't allow us to produce enough food, as grains both struggle to grow and grow with fewer nutrients as heat levels rise. While some new land will become arable in the polar regions, it's not as rich, and it won't make up for the losses. We will, as the scientists interviewed for the article put it, be trying to feed more people with 50 percent less food.

The potential impacts continue. There's a risk of what the article calls "rolling death smogs" of air low in oxygen but filled with lung-busting levels of particulates from drying, burning forests. The decrease in air quality will harm the elderly, the young, and those with compromised respiratory systems. For the rest of us, lower levels of oxygen will make it harder for us to think, and levels of illness will skyrocket, as they have now in critically polluted China.

There's the impact of devastating, constant storms and fires on our industrial capacity, which will be pressed by breakdowns in transportation infrastructure and logistics. A study of economic impacts published by Stanford University found that, in the near term, the resulting impacts would ding our economy with a 20 percent or more drop in global production. This would essentially result in a permanent and irreversible Great Depression, one that would involve the vast spread of poverty.

With rising sea levels and the accelerating melting of the polar ice caps, we'd see the inundation of all coastal cities, communities, and facilities. Most of us may have written off New Orleans with a sigh. Many of us may have assumed that Miami isn't viable long term and that the lovely beaches of North Carolina's Outer Banks will no longer be around in fifty years. But when you map the time horizon out further, to a hundred years, and then two hundred, the impact becomes more substantial. The face of the earth changes. It's not just that our descendants five generations hence won't get to have a margarita on Key West as they watch the sunset with Jimmy Buffett playing in the background. They won't even be able to travel to Florida because Florida—pretty much the whole state—will no longer exist. That's hundreds of years away. But it is still a distinct possibility.

With rising sea levels, decreasing crop yields, and increasing heat, the human beings who live in the most heavily impacted areas will need to go somewhere. The images we see now of refugees arriving, desperate and empty-handed? That will be magnified exponentially, with the mass exodus of not hundreds of thousands but hundreds of millions. And with these migrations and sudden resource scarcities, we can expect war upon war as human beings fight over remaining arable land and resources.

As if that's not bad enough, there's the real possibility of plagues as disease vectors from the equatorial regions move into the formerly cold north and ancient viruses locked into the permafrost are released into a human population that has no developed resistance. Global pandemics, as we now know, aren't all that much fun.

But wait—there's more. If this cycle continues to progress in the way of other, similar events in the long history of our planet, there's a nonzero probability of the extinction of 97 percent of current species, up to and potentially including all of us.

Cheery stuff.

Some of this, God willing, may not happen. None of it is entirely certain. But much of it, or some variant of it, is likely.

It won't happen all at once. It's not like an asteroid strike or a nuclear war between major powers. You do not wake up one morning, make a cup of coffee, and then . . . boom . . . climate change! That, as I argued a couple of chapters back, simply isn't how this is playing out. That's one of the primary reasons that global warming is so difficult to mobilize against. At the time of this writing, folks are starting to try to rebrand climate change as a "climate emergency," or a "climate crisis," in the same way that there was a rebranding when "the greenhouse effect" or "global warming" just didn't seem to have purchase. It's the kind of emergency where most days feel the way they did the year before in most places. Instilling urgency just doesn't work when things don't feel urgent. There is no greenish-black sky, no freight-train roar of a funnel cloud bearing down as the warning sirens wail.

In fact, it is most likely to play out over several hundred years, with longer-term impacts lingering for hundreds or thousands of years. Which, on the one hand, means there's time to prepare. There is time to adapt, even as these eventualities arise.

On the other hand, it means that we as a species will be stuck on a harsh, merciless world for a period of time exceeding the entirety of our written history. This surfaces a strange truth: when we describe the extreme weather events caused by climate change as "historic," we're not quite getting at it. Human history only extends back a handful of millennia. The events that are coming our way haven't been seen on this planet for millions of years. Our shallow memory as a species does not prepare us for them. Even the five thousand years between us and the first writings of the Hebrew people don't reflect enough experience to have a meaningful frame of reference.

It feels completely overwhelming. It's really, really tempting to just throw one's hands up in despair or curl up into a fetal ball in some nice, dark corner of one's Instagram account and look at puppy videos until the end comes.

But apathy, cynicism, and despair are all maladaptive responses to a crisis that will test all of humankind. Enduring, surviving, and maintaining personal and community integrity in crisis require that we look at the chaos and the threat and the coming-apart-of-things not with fear but with a deep conviction that we can overcome. That's not easy, I know.

It's just too much. We just can't even, as the saying goes. Existence, even at the most humble of levels, can often feel overwhelming.

A few years ago, I can recall having just that sort of overwhelming moment. It was a weekday, and the work of the day was done. I was planning on getting to bed at a Ben Franklin–approved healthy, wealthy, and wise hour. I was going to settle into my comfy dad-chair and read a novel until fatigue took me, which I did. "Hey, I'm going to bed," I said and turned in as the rest of the family still puttered about the house. It was nice and neat and just so, an evening that followed the natural order of household evenings.

But just as I reached REM sleep, our dog started having a seizure, threw up, pooped, and fell down the stairs. I woke up and, my mind still cloudy, stumbled out to clean up the mess and figure out whether she was injured. She was struggling to walk, and it looked like an emergency trip to the vet was necessary. And right as we were trying to deal with that crisis, my older son announced that he was starting to feel lousy, and lo and behold, he was running a fever near 104 degrees, his eyes bleary. Ack, we went, running around in a parental panic as our orderly expectations came apart.

Things had collapsed, going from order to chaos, in less than ten minutes.

Existence, or so we are told, bends toward disintegration. Chaos is, we hear from some corners of science, the very state and nature of the universe. Order degrades, and all descends to entropy. Things fall apart, as Nobel Prize–winning Nigerian novelist Chinua Achebe reminded us. The universe is slowly, surely, declining, as columnist Michael Gerson wrote in a particularly reflective op-ed.

These things are true and feel ever truer as the years progress. Few things remind you more of the gradual degeneration from order to disorder than your arrival at midlife, as your body aches and sags and randomly decides to go out on you.

Yet in the face of that, there's the reality of life. Not my own life, but life itself, as we can observe it. Life seems to drive fiercely and intentionally in the opposite direction. Life moves from complexity to complexity, growing ever deeper and more sophisticated as it grows and evolves and adapts. From random bits of protein to cells to multicellular organisms to social organisms, from the flail-around-until-a-mutation-sticks adaptive spamming of evolution to the intentionality of sentience, life shows a peculiar trend toward more and more elegant systems as it tacks hard through the waters and winds of chaos.

Life moves against the flow of the second law of thermodynamics, in ways that appear to be nonrandom. It is possible, I suppose, to consider sentient life as an anomaly, just a swirling meaningless eddy in the great current of entropy.

I believe that it is something more—something that must be part of the system, that is spoken into being by the One who is the source and completeness of all being.

That belief gives me a powerful sense of purpose, even in those times when things have fallen apart, even when the light of hope glimmers faint and far off. Faith does many things for a soul, but perhaps the most potent and vital gift of faith is that it gives us a way to cope with events that would otherwise tear us apart or break us.

There are many such frameworks of faith, but the one I am particularly interested in—you know, being a pastor and all—is the one that Jesus of Nazareth teaches. As we cross over into a time of chaos, that faith can be a source of strength, endurance, and guidance for us.

What, then, does that look like? How will Christian faith embolden us to forge on and find our way on a planet that has turned in anger against us? What can our faith say as this impossibly vast crisis rises to meet us?

CHAPTER 9

CHRISTIAN MORALITY AND CRISIS

Funny word, *crisis* is. We think of it as a simple synonym for catastrophe—just another way of saying the word *disaster*. The word's root meaning goes far deeper, though. It derives from a Greek word, κρῖσῖς, which means "a time that forces a decision" or "a time of judgment." Crisis is all about our choosing and showing who we really are.

In Scripture, we encounter crisis repeatedly in the teachings of Jesus. In John's Gospel, for example, we hear, "And this is the *krisis*, that the light has come into the world, and people loved darkness rather than light because their deeds were evil" (John 3:19). Or this: "Now is the *krisis* of this world; now the ruler of this world will be driven out" (John 12:31). Or this, from Matthew's Gospel: "I tell you, on the day of *krisis* you will have to give an account for every careless word you utter" (Matthew 12:36).

Understood from the teachings of Jesus, a crisis is a test of our authenticity as souls. Do we hold to our principles, or do we set them aside? There is always the temptation to set principles aside when the going gets hard. That is a challenge for us as citizens of this country and inhabitants of this planet. This challenge—holding on to virtue when things get tough—will face Christians as this event we have inflicted on ourselves unfolds over generations.

Our lives and our culture? They offer up tests, moments of *krisis*, every single day. Power and pride, anxiety and greed, desire and self-serving—all of these things come pouring toward us. When they do—and when they try to command your allegiance and your obedience—remember the words of Paul: "Test everything; hold fast to what is good; abstain from every form of evil" (1 Thessalonians 5:21–22).

This is true in every instant, in every choosing, and in every action you take. When you feel yourself stirred to anger by whatever overhyped thing everyone is angry about today on Twitter and you feel—in that moment—that you *must* weigh in on something that you didn't even care about five minutes ago? You test that urge. Do you need to be hostile toward that complete stranger? Do you need to snarl back at the rude comment an equally complete stranger directed at you?

No. No, you do not. You stand fast in what is good. You maintain your integrity, and you either hold your peace or speak from a place of grace.

Or let's say, for a moment, that you are a married pastor in your late thirties, harried and rumpled and stressed by life and the demands of church and family. You're at church on a Sunday afternoon, and during a conversation, a young and attractive parishioner—whom you know, from counseling sessions, is struggling in their marriage—makes an overtly flirtatious comment to you. There are a number of different ways you could respond. You could, from that place of feeling both complimented and a little attracted, flirt back. You could edge closer to that place of transgressive energy, enjoying its danger and forbiddenness. You could hold that moment as a secret pleasure, something to revisit in your mind and taste again more deeply.

Or you could test that temptation against the future it might create, against what it does to your soul and to your covenant commitment to your spouse. When you get home, you could make a point of telling your spouse about that parishioner's forwardness, burning away the temptation of secret transgression in the light of your vows to the one you love. You could also mention the exchange to a trusted elder of the church, not for the purposes of shaming that parishioner, but to set up a hedge of accountability around your integrity and your calling as a pastor.

Not, of course, that I speak from experience on that one. That's entirely hypothetical.

For Christians who hold their commitment to Jesus of Nazareth as the heart of their integrity, every moment of choosing—from the seemingly trivial to the monumental—is a *krisis*. And in those moments, we have the witness of the Gospels to guide us in our actions and the teachings of the Epistles to show us the path of Christ's grace.

As we approach a great, multigenerational time of testing, Jesus has never been more relevant to our moral choosing. The crisis of climate change is a moral crisis, one that will determine whether our commitment to the heart of the gospel of Jesus Christ is real or just window dressing. It's going to require us Jesus-folk to test our beliefs and be sure that we haven't replaced the fierce moral demands of Jesus with heady academic abstraction, political ideology, or simple grasping self-absorption.

My own ethical responses to the climate crisis are driven by many things. Chief among them is this rather fundamental understanding of Christian faith: there can be no difference between my faith and my actions in the world. If I say I am committed to Jesus, and to being guided by his teachings, it is that obedience in word and deed that defines the legitimacy of my faith. Period.

This is not to say I'm at all perfect in that regard. I'm not. I screw up. I act unfairly. I let bitter words fall from my mouth. I feel avarice, particularly where fast cars or motorcycles are involved. I "lust in my heart," as dear brother Jimmy Carter once put it. That's why I'm so very thankful that grace and forgiveness are at the living heart of the gospel.

But the grace and forgiveness of the gospel are not an excuse for moral sloth. If I do not in good faith seek to discipline my actions and my choices to conform them to the specific ethical demands of Jesus, then I'm not a disciple. In that, I follow the blunt, earthy, uncompromising faith of one of the key influences in my Christian journey: George MacDonald.

MacDonald was a Scotsman, a mystic, and the spiritual teacher of C. S. Lewis. Like most true mystics, MacDonald's faith wasn't abstract. It was fundamentally practical, recognizing that the false binary between faith and works in Christian theological blabbering is meaningless. As Mac-Donald put it, belief just isn't real unless it impacts how you act: "Instead of asking yourself whether you believe or not, ask yourself whether you have, this day, done one thing because He said, Do it! or once abstained because He said, Do not do it! It is simply absurd to say you believe, or even want to believe, in Him, if you do not do anything He tells you."

Being a disciple of Jesus of Nazareth is no more and no less complicated than that. Saying "Jesus is my Lord and Savior" is all well and good, but Jesus ain't your "Lord" if you ignore everything he tells you to do.

Faced with the crisis of the gospel, it is deeply tempting to contextualize away our obligations. The moral demands of Jesus, many Christians say these days, are just one choice among many. Who's to say what's right or wrong, really? Who am I to invalidate the choices of others? True enough: someone else's decision is not mine to judge. But that doesn't mean that, as a Christian, one can believe in moral ambiguity or that our decisions don't matter.

There is no basis in the teachings of Jesus for assuming that. What we do matters. How we act in this fleeting life of ours has eternal weight and significance. Some Christians want to avoid this, as it smacks of being judgmental and manipulative. They've had enough of hellfire and brimstone talk and would rather just present a God who never judges or a God who only exists to pour out material blessings on the faithful. "None of this hell talk," they say. "It's just mean."

Let's look at that for a moment. I've often heard H-E-Double-Toothpicks described as the absence of God and God's presence. Hell, so we sometimes hear, is an existence in the absence of God and God's love. Hatred and acts of evil cannot stand before God; therefore, they cease to have a part in God's eternity, glowering forever in the shadows. On many levels, I resonate with that position. It's gracious and has some scriptural support.

On the other hand, I do not believe that anything can exist or does exist in the absence of God's creative power. Outside of God, there is only a void, emptiness, or formless chaos. If hell is understood as simple nothingness, it is closer to the ancient Hebrew concept of Sheol: an absence of existence.

To be honest, I'm not sure that simple oblivion—the negation of personal existence—captures the depth of the reckoning to which Scripture bears witness. If there is to be justice, would it be served by simply erasing Pol Pot or Osama? Is that "reaping what you sow"? Perhaps . . . but I think there are other theological alternatives.

The primary theological challenge, it would seem, is reconciling eternal damnation with the Christian affirmation that God is love. How can a loving God be wrathful?

To that end, it helps to have an idea of what love is. Numerous Christian writers have dealt with this—C. S. Lewis's *Four Loves* is an excellent primer. Most theologians hold that God's love is *agape* love: a love that transcends self-interest or emotionalism and involves full participation in the life and spirit of another. Being a flagrant theologeek, I tend to favor the articulation found in the writings of twentieth-century existentialist theologian Paul Tillich. Tillich defines love as our yearning to participate in another being, to truly know that other being, and to share in their joys and their pains. Love is seeking yourself in another, reaching out across the boundaries that limit us. Love is our struggle against our separateness, our struggle against all that divides us from one another. We human beings do this clumsily and imperfectly, if at all.

But while we struggle to make love real, God exists as love in its perfection. Love is God's very nature, and love is, therefore, the foundation and root of all existence. We are created in love, and to love we will return. When we profess a faith in a God who *is* love, we profess faith in a Love that tears down the boundaries that divide us and in a Love that allows us to truly be a part of another.

So if God is love, what is the judgment of a loving God?

For the comforters, the gentle, the humble, the peacemakers, the seekers of justice and equity, the answer is that the measure of God's justice is love. The love of God will open the servants of love to every being they have touched. Those who have lived their lives in compassion will know the fruits of their actions as their own. The walls that separate them from the other children of God will fall before the love of God, and in the fullness of God's love, they will participate fully in the lives of everyone they have touched. Each and every life that they have touched, each and every relationship in which they stand, will be their heritage for all eternity.

But what of the oppressor, the destroyer, the self-seeker? What of the one who seeks gain at the loss of others? What of the one who rules with an iron hand? What of the ones whose hate drives a bomb-laden car into a crowd of innocents, leaving death and terror in their wake? The one whose willful ignorance brings others to harm, or the one whose actions contribute to the suffering of others?

The answer, again, is that the measure of God's justice is love. The love of God will open them to every being they have touched. They will know the fruits of their actions—or their stubborn willful inaction—as their own. The walls that separate them from the other children of God will fall before the love of God, and in the fullness of God's love, they will participate fully in the lives of everyone they have touched. Each and every life that they have touched, each and every relationship in which they stand, will be their heritage for all eternity.

Those hellfire-and-brimstone preachers have it all wrong. Why preach of wrath? Why preach of wrath at all?

There is nothing more terrifying, more inescapable, and more purely just than the love of God. This is the heart of the gospel crisis and the understanding that must drive all authentic Christian choosing.

From this foundation, the Gospels and Epistles have some very specific things to say about Christian moral obligation. If we are to stand in the face of a loving God whose only measure of our worth is love, that means we must take our moments of choosing—those crisis moments—and be guided by those teachings. Building on the witness of the Hebrew scriptures, the moral teachings of Christian faith present us with a set of universal values for life that haven't changed over thousands of years and—because we are still just as human—are just as relevant today as when they were first written.

And these values are even more deeply vital for our decisions as we move into an era of radical crisis in which our ethical commitments will be tested as never before. Where the church is present after climate change, bearing witness to Jesus in that hard, harsh world, it will need to respond to this moral crisis in some very particular ways. In order to do that, though, something more is needed.

What's needed is the conviction that our actions will make a difference and that they serve a purpose. What's needed is a sense of hope.

CHAPTER 10

APATHY, APOCALYPSE, AND MORAL ACTION

If things are as bad as they seem, the question begs itself: Why act at all? We're doomed. Nothing can be done. It's all over. The temptation is strong to simply continue with business as usual because, really, why not?

Oddly, this impetus to inaction seems to rise from two diametrically opposed streams of thought. On the one hand, you have those who obsess over every last bit of negative news about the climate. Every drought, every out-of-control wildfire, every storm is a sign that things are past saving, that the great steamroller of planetary ecological collapse is coming for us all. Best just not to breed—we should be thankful that we won't personally live long enough to see the world burn.

On the other hand, we have the Christians who are convinced that none of this matters because all the signs point to the return of Jesus at some point really soon. An entire cottage industry of cable-television Bible prognosticators pores over every item of news as intently as a druid sorts through the entrails of a butchered goat. The storms and tumult upon the earth must mean *something*.

Christians have been sure Jesus is returning any minute now for two millennia, but this is no impediment to the John Hagees and Jack Van Impes and Harold Campings of the world. As we all know, God created everything so that we could be at the very center of it, fashioning all of time and space for us right now. Not the generations past. Not the generations to come. But for us.

Oy. If that's true—the world is coming to an end anyway—why should we be bothered to act? Why bother to save it at all if it's just something God will be destroying in the very near future? We're all just going to be raptured away, right?

The Rapture—this idea that before the going gets hard, the faithful will be swept up into heaven—has inexplicably become a central part of Evangelical doctrine. Folks eat it up with a spoon. Tim LaHaye's narratives of the Rapture have sold an abundant pantload of Left Behind novels and produced some of the most ragingly unwatchable films in the history of moviemaking.

Honestly, I have never gotten it. Not even a little bit. I know it draws inspiration from an interpretation of one section of Luke's Gospel (Luke 17:34–36), in which Jesus says some are taken and others are left. And . . . well, that's pretty much it. There are some extrapolations, followed by some interpretive gyrations, followed by some Olympic-level proof texting, but it's essentially just that one little chunk of text interpreted through the warping lens of the Archangel Scottie and his Bible-believing transporter room. I know many folks believe this, taking it on faith that the message of Jesus includes an escape clause for the righteous. None of us want to suffer. I get that. If we've done right, we hope and trust that it will matter to our Creator.

But here's the essence of my problem. On a recent morning, as my sitting-in-traffic mind immersed itself in a section of Faure's Requiem Mass that was pouring through my six-speaker sound system, I found myself in a Rapture reverie. I found myself viscerally envisioning that moment, were it to happen to all of us.

Now, I know this is a stretch. As a liberalish kind of Christian, I know I'm not really on the Tim LaHaye short list of folks likely to get taken up. Still, one never knows.

So there I was, envisioning what that moment would be like. The world is coming apart. Earthquakes. Fires. Buildings crumbling. People crying out in terror. And as one chosen, I'm unaffected. I'm rising up, not really bodily but into that deeper reality of God's presence. I'm suffused with light, radiant with the power of my ascension to a place of peace and glory, my physical form yielding to my spiritual body as I began to move beyond the spreading cries and conflagration.

It would just suck. I'd feel horrible. It would be the worst moment of my soon-to-be-over corporeal existence.

Why? Because Jesus matters to me. What he taught and how he lived matters to me. His radical and transforming compassion matters to me. And so in that Rapture scenario, I would look down at the fading, burning world, and I would weep. I'd want none of that suffering for any of those souls remaining, even those who have hurt me deeply. I'd feel not satisfied, relieved, or joyous but consumed with horror, loss, and disappointment.

I think that's how Christ would have felt if, from the cross, he had seen the story play out differently. What if, crying out from the cross to God to forgive those who were murdering him, Jesus had watched the heavens open and this sinful world be consumed by annihilating fire? What if he had watched the fury of a father destroying everything that had hurt his child? I can understand that anger, but it bears no resemblance to Christ. It's a human rage.

If the fire had consumed centurions and swept aside zealots and Pharisees, Christ would have seen it as betrayal. For Jesus, watching the world burn would have meant the burning away of the future of his message and the burning away of the possibility of us. He would truly have been forsaken. His purpose, all his love, all his hope, all his teachings, all his transforming *Logos*? They would have been wasted.

The reason the Rapture works for folks, theologically, is that it is all about us. It says, in defiance of the cross, that real Christians don't have to suffer when the world falls apart. It says, in defiance of our incarnate Lord, that this world doesn't really matter. It reinforces ego and a sense of otherness at the expense of the hard kingdom compassion that lies at the heart of the gospel. It just isn't Christian. Many Christians believe it, but it bears no resemblance to the essence of what Jesus taught. So it bugs me.

But it bugs me for another reason. I'm not sure there's a way to put it nicely, so I won't. That way of viewing God's work is shallow and selfish, devoid of love and utterly compassionless.

When we use self-centered, end-times speculation to justify inaction, it's also . . . it's . . . hmmm. Let me use an illustration drawn from my own long history of doing dumb things. Years ago, I was rushing to a job interview. I was stressed out, running late, and anxious about getting there on time. As I live inside the Beltway, there was a ton more traffic than I'd expected, and I had miscalculated the time required to get to my destination.

Up ahead, a construction site had reduced four lanes to one, and I sat and grew increasingly frustrated at the trickle of cars they were letting by. The little glowing green LCD clock on the dashboard of my old Honda showed that I only had about five minutes to get to where I was going. The flagger waved us through in a painful dribble. Five cars. Then another five. I inched forward.

When my little group was finally given the go-ahead, I gunned it. I also looked at the man holding the flag, my heart filled with impatient annoyance. I glared at him with my most ferocious glare as he waved me by.

Then I looked where I was going.

I was heading straight for the back of a BMW, stopped at the red light just thirty yards past the construction site. I had not seen the light. I'd not been looking where I was going. Given how much speed I'd picked up and my irritated inattention, there was also absolutely no way I was going to be able to stop in time. Impact was inevitable. It took me a fraction of a second to process this.

What was my next action?

Given that God's time is not our time, what should my choice have been? An eon in God's sight is no different from a moment, and the blink of an eye is of no more import than the whole span of human history. If I assume that the return of Jesus is imminent—that it will occur at any moment and that it will arrive just as I, the Chosen One, am in peril—why should I have braked at all?

"Jesus, take the wheel," I could have said, closing my eyes and spreading my arms.

Theologically, that is what countless Christians argue when faced with the climate crisis. We need do nothing. Jesus will take us before the crash comes. There's really no difference, seen through the eyes of faith.

Both are equally wrong.

When a moment of disaster is about to strike, you act. Even if you realize you won't escape, you do everything you can to mitigate the damage. Anything else is morbid, antilife, self-annihilating madness.

Even knowing I couldn't stop, I did what you would have done. I slammed on the brakes as hard as I could, pulsing them to maximize traction as they locked up in my pre-antilock-era Honda. I plowed into the back of that Beemer, sure, the nose of my Accord diving into the rear bumper. The bumper shattered and crumpled the front of my car.

I leaped out, rushing to see how the other driver was doing. She was a young woman, and she was really upset—but not about the car. She was in a panic, her eyes brimming with tears as she turned toward the back seat, saying a girl's name over and over. I followed her gaze to where her baby girl sat completely silent and still in a car seat.

There was a moment of horror.

Then the kid, getting over her long startle at the impact, started squalling.

The baby was fine. The mom was fine. My car was trashed; hers not so much—although it's amazing what a few scrapes to a BMW will set you back. My insurance rates justifiably went up, but that was about it. I had gotten on the brakes as hard as I could, even knowing there was no way I'd stop in time. It could have been so much worse.

In the face of disaster, healthy moral souls don't just give in or assume that some miraculous intervention will get them out of it. They act in ways that increase the probability of the best possible outcome. That action is not in defiance of faith. It is the fruit of an authentic faith.

Christians have always struggled with that dynamic. We struggle to find the balance between trusting in God to provide and guide and the need for us to act on the ethical and moral teachings of Jesus. The argument, or so we've allowed ourselves to imagine it, is between faith and works.

This is and has always been silly. Belief and action are woven up together inseparably.

When Christians talk about belief and action, there's no better place to start than with the book of James. James is all about seeing and being and doing. This "letter" toward the end of the New Testament is more an essay, or sequence of essays, that establishes the most essential nature of the Christian ethical life. It has been traditionally attributed to James, the brother of Jesus. As it is perhaps the most theologically Jewish of the books of the New Testament, most contemporary biblical scholars see no reason to challenge that.

James is one of the most practical, rubber-meets-the-road books in the New Testament because it is essentially a book of wisdom. As Wisdom literature, it concerns itself with how we human beings should act if we're to get along in the world. Wisdom teachings are found woven throughout the witness of the Bible: in Proverbs, in Ecclesiastes, in Job, and in a number of the Psalms. These writings teach the basics of how to live—and particularly how to live so that you are playing well with others and doing well for yourself. These books teach thrift, foresight, and patience. They teach that life is to be enjoyed and that sustained enjoyment is best found in moderation. They teach that the wise do not speak without careful self-examination, first considering the impact of their words. They also teach about the balance between human power and human liberty.

As a teacher in the wisdom tradition, James is as straightforward in his words about what is good as a two-by-four applied directly to your forehead. In a nutshell, goodness is defined as actual, material, and physical obedience to what James describes as the "royal law." What is good is loving your neighbor as yourself. Period. By that, James does not mean love as an abstraction or love as some distant, squishy, ethereal concept. As James puts it, "What good is it, my brothers and sisters, if you say you have faith but do not have works? Can faith save you? If a brother or sister is naked and lacks daily food, and one of you says to them, 'Go in peace; keep warm and eat your fill,' and yet you do not supply their bodily needs, what is the good of that? So faith by itself, if it has no works, is dead" (James 2:14–17).

Throughout the letter of James, what is most significant is a compassion that directly acts and a depth of valuing other human beings that shows no partiality. This is, for James, the essence of what it means to be good.

We Jesus-folk have let ourselves fall into the trap of imagining that James is somehow opposed to Paul, a trap that may or may not have been sprung by dear ol' Martin Luther during the Reformation. Luther, who rightly felt that he could never, ever be perfect enough, saw in James a demand for unattainable perfection. In the context of his struggle with a church hierarchy that argued that "right actions" could be monetized and that heaven could be purchased, Luther snarled back against James. It was, he said, "an epistle of straw."

But that vision, born in the crucible of radicalizing conflict within the church, just doesn't represent how vital it is to live out of our moral commitments. It creates an imagined tension between James and the apostle Paul that simply doesn't exist.

From Paul's writings, and particularly his letter to the church at Rome, we Christians assert that we are saved by grace and not by works. Therefore, someone who does good but has not proclaimed Jesus Christ as their Lord and Savior is just doing "works righteousness"—so the argument goes—which is pointless and worthless.

I've always had a beef with this, truth be told. Why? Because under this rubric, there is no difference between an "unbeliever" finding you injured in a wreck, stopping, bandaging your wounds, and getting you to the hospital on the one hand and that same "unbeliever" finding you injured in a wreck, taking your wallet and shoes, punching you repeatedly, and then slitting your throat on the other. From the perspective of the Evangelical movement, any distinction between these acts is meaningless to God. Both are equally evil, for the person undertaking them is automatically damned no matter what they think or how they act.

What's most difficult about this for me—beyond its self-evident disconnect from the idea of "good news"—is that it radically misrepresents Paul's essential point about works, faith, and righteousness. What are "works"? Well, they're anything you do. Anything. Building a Habitat for Humanity house? That's a work. Popping a cap into some fool who disrespected you? That's a work.

What is the apostle Paul talking about when he describes "works" that do not save? Random actions? Any actions? Evil actions? No. The "works" being challenged are works "under the law" (Romans 3:19–20). What Paul is challenging is the idea that obedience to an external code of conduct—in this case, the Torah—has any power to restore our relationship with God. Why?

Because law and legality assume an underlying enforcement through coercion. It's how the state runs. In the contract between a ruler and their people, failure to comply with the terms of a social compact will result in unpleasantness for those who mess up. That ranges from small fines to more unpleasant things, particularly if you live in the district of Sen. Vlad Dracul (R-Transylvania).

But if that's the reason you engage in moral action, it means that you are beholden to fear. You are not acting as one moved by Christ's grace, meaning you are not inwardly conformed to God's will through the action of the Spirit. You're just doing what you're told. It's a shallow, meaningless, untransformed obedience, rooted in a terror of divine punishment.

Paul was proclaiming that through Christ, that whole dynamic was shattered and replaced with an awareness of God's grace. Not God the divine autocrat, but God who moves to change our hearts to the good with a relentless and inexorable grace.

Yet when we see individuals who are not law driven, when we experience souls who seem driven to show care for others not because of fear of

God but by some deep upwelling grace, for some reason we feel compelled to declare them damned by the name of Jesus.

In the name of grace, a sizable percentage of Christians are willing to be graceless. To fairly paraphrase Paul, "You who brag about grace, do you dishonor God by showing no grace? As it is written, 'God's name is blasphemed among the Gentiles because of you'" (Romans 2:23–24).

If we claim to be Christians and our whole selves are not oriented to the radical, transforming teachings of Jesus, then we ain't gettin' it right. It's not a question of demanding the impossible perfection that Luther imposed on himself. It's a question of acting.

It is not up to Jesus to pay attention for us, or to steer for us, or to hit the brakes when we're heading for disaster.

Pay attention, he says. Turn, he says. Stop, he says. (See Matthew 25:1–13; 4:17; John 8:11.) Our task is to listen to Jesus and, as the body of Christ, respond by doing what he tells us.

As this time of global crisis unfolds, it matters how we respond—not just in word and belief but in action. If we are to maintain our moral integrity, both heart and hands need to be turned to the work of enduring the impacts of climate change together. In the chapters to follow, we'll look at some of those impacts. We'll examine just how our commitment to follow Christ in all things must, if we take discipleship seriously, shape our moral action in the face of a global crisis.

PART IV

FOLLOWING JESUS AFTER CLIMATE CHANGE

CHAPTER 11

ADAPT TO DIASPORA

One of the more fascinating things about serving a small church in a small country town has been encountering a community that has a sense of history. That goes beyond discovering hundred-year-old bottles of Guinness stuffed into the walls of the church building where your office is located. It's a personal thing. It's a community thing. There are families that have been here for generations, and that's different from what I'm used to.

That is in large part because I'm "from" a place that almost no one is actually from. When you live inside the Beltway—in a suburban moonscape that sprawls as far as the eye can see—you're in a place where pretty much everyone comes from somewhere else.

I'm not from that small town where my little community worships. I'm inside that Beltway. My hometown of Annandale, in contrast, is a place without evident history. Folks don't go back that far here because the *here* that is there now wasn't there a hundred years ago. It was a few farms, a few estates, and a whole bunch of woodland. There was a little Methodist church at a crossroads, one that had the misfortune of being burned to the ground during the Civil War. Annandale was more of a place than a town.

It stayed that way until a great tide of ramblers and asphalt swept across the land in the late 1950s and early 1960s, and suddenly there were people there. Whatever sense of history existed in Annandale now rests under strata of asphalt, tract homes, and stubbornly struggling Korean businesses. Here and there you'll find a sign or a historical marker, sitting rusting on a patch of pavement by the Walgreens.

There used to be something here, that sign will say. But it isn't here anymore.

I think that's the Annandale community motto.

You can't really be rooted in a place that has no living roots. You're just another transient, drifting through for a year or two before your next assignment or your next posting or the next job calling you on to another place. Everyone is a stranger, and just about when they stop being a stranger, they move. We've been ensconced in our little slice of suburban paradise for just over thirteen years. Our neighbors to the right? Different. Our neighbor to the left? Different. The guy across the street? Different. We're starting to become the old-timers, and we've been there only a half of a generation.

Annandale's most famous resident is sort of a poster boy for our transience. Who is that? It's Mark Hamill. Yes, Luke Skywalker himself lived in Annandale and went to Annandale High School, but for Mark, his time in Annandale was as fleeting as Luke's time on Tatooine. He lived there for a few years, but his dad was a military officer, and Mark Hamill found himself having to move around a whole bunch. Having to constantly move because of your father's job was yet another reason Mark Hamill was the right person for his role in *Star Wars*, I think.

Annandale is increasingly the face of America. It's what everyone's lives are like, everywhere in the United States. Americans, generally and as a rule, come from somewhere else. Almost no one is "from here."

Transience is a basic part of our culture. We're forever churning and cycling, chasing after one thing or another. Our society roils relentlessly, like ripples on the surface of a bathtub, bouncing from coast to coast now that we've filled up our allotted portion of the North American continent.

And as much as we like to tell ourselves that our compulsive transience doesn't take a toll on our souls, it troubles us nonetheless. Moving unsettles us and leaves us feeling disconnected and disrupted and like we're not quite sure of our place in the world, even when we choose it.

It's an experience familiar to Indigenous American peoples, whose encounter with the wave of colonization and immigration from teeming, overstuffed Europe involved repeated and sustained loss of place. A century of forced relocation impacted identity and self as peoples were removed from their ancestral lands. In her book *Braiding Sweetgrass*, botanist Robin Wall Kimmerer tells story after story of the struggles with self that can come from a loss of place. Her Potawatomi people were forcibly moved several times, and each time, there was a struggle for memory and self. As she describes it, "In the span of a single generation my ancestors were 'removed' three times—Wisconsin to Kansas, points in between, and then to Oklahoma. I wonder if they looked back for a last glimpse of the lakes, glimmering like a mirage. Did they touch the trees in remembrance as they became fewer and fewer, until there was only grass? So much was scattered and left along that trail. Graves of half the people. Language. Knowledge. Names."

Though the strange, churning suburban identity of America's eternal diaspora seems far removed from the visceral twining of place and community that Kimmerer describes, she has called out something that speaks powerfully to most human beings. Our context shapes us, even if we aren't conscious of it. We are creatures formed by our connection to the land.

The great challenge that will face us as we live in this land that stretches from sea to shining sea is that those shining seas are going to be closing in on us a bit over the next several hundred years. Even if we slow our production of emissions, most projections of sea level rise point to a fundamental reworking of our coastline as communities that have been happily thriving for a century or more find themselves forced into retreat by rising sea levels.

My own recent family story includes just such a place. Back in the 1930s, my paternal grandparents built three houses on Fire Island. It's a barrier island that sits across the Great South Bay from Long Island. Fire Island is a spit of sand thirty miles long and just two blocks wide. Ninety years ago, Fire Island was barely populated, and my grandparents decided—rightly—that it was a great place to build and invest. It was and is a magical place, a place without cars. There were "roads" on which a single emergency vehicle could travel. But private vehicles were verboten. That meant that the island was the kind of community where kids could amble about blissfully and completely free. Fire Island meant the sand between my toes, lollygagging about, offering up day after lazy summer day. I'd go to the little library on my scooter. I'd wander to the arcade. I'd find other neighborhood kids and we'd have a squirt gun battle for hours, with nary a hovering adult in sight. Those were the same childhood memories my father had as a preacher's kid growing up on an island where a kid can run free. (There's a delightful book called *Pinky Pye*, which follows a family living on Fire Island and the cat that adopts them. It's perfect bedtime storytelling.)

Fire Island is also barely above sea level. I've ridden out Cat 1 hurricanes on that island and watched as waves from the bay lapped up against the house.

Years and years ago, as my grandparents retired, they offered the last of the houses to my dad. It was tempting, particularly given that Fire Island is (1) magical and (2) really close to New York City. Item two means that property there ain't cheap now. An idyllic beach retreat convenient to Manhattan? Lord, but that'll cost you. Right now, you can only get monthly rentals there, and they wander up into the six figures. For a month. My grandparents were very, very right about the value of that investment.

But my father demurred, mostly because we didn't live in New York, and maintaining a property that could at any moment be swept into the sea seemed not to be the best bet. The family sold off those properties.

Back in the 1980s, that was a hard choice. Now? It seems like the right call. Fire Island, though it's a place of beauty and community, is doomed. Even before sea level rise, it was right on the cusp of viability. When you add three, four, six inches to the water level? Places like Fire Island cease to be habitable.

Longer term, as we look to the next century, that reality will press in deeper. It's not just little barrier islands and vacation resorts that will be challenged. It's places like Key West. Miami. New Orleans. Or Bethany Beach, where I and my family have vacationed since before my adult sons were born. Major cities and heavily populated areas all along the world's coastlines will be compromised to the point of unlivability. Storms and rising waters will take deeper and deeper tolls, to the point where rebuilding will become less and less tenable.

This is not going to be easy for us. We're territorial creatures, creatures of place, for whom giving up a familiar home or neighborhood is like giving up a part of our identity. Those places are woven into our memories, filled with recollections of children and parents and grandparents, of first loves and laughter and loss. Being forced away from our places will feel like a loss of self.

For most of us, this is a difficult thing. Even for Americans, in our rush and rootlessness, it's hard. It is the place where we become anxious and angry. It is a place of desperation, one that normal human beings hope never to encounter.

If we reach back into Scripture, we find many stories of normal human beings forced from their land—people like the prophet Ezekiel. Ezekiel, though, was not a normal human being. Like many prophets, Ezekiel was a weird, weird dude.

He needed to be weird because he lived in weird times. Ole Zeke was a member of the priesthood with a clear career track as a priest in the temple in Jerusalem. That was his whole life and the span of his family history. He was a Zadokite, meaning the priesthood was in his blood. His daddy had been a priest, and so had his grandpappy.

As he'd grown up, he would have learned all the rituals: all the complicated prayers and songs and sacrificial techniques. He would have learned

how to dress, how to eat, and how to follow the law to keep himself pure
for the temple. It was his whole life, his whole understanding of himself.

And then that thing was destroyed. Ezekiel shared his visions with
a people who had watched Jerusalem burn. They had seen their leaders
butchered. They had been driven from the land that had belonged to their
ancestors for generations stretching back into legend. The temple—the
holy of holies, the footstool of the God of Israel—that temple lay in ruins,
battered and burned, all of its sacred and holy objects looted or scattered
to the four winds.

Ezekiel's people were lost. Their hopes were dead. Their future was a
dead thing, shattered by the military might of Babylon and blowing like
ashes throughout the empire. Ezekiel himself had been taken with them,
and he found suddenly that all the tools of his trade, all of the rituals that
made him a priest over the people, were now meaningless. What does it
matter if you know how to do a perfect temple worship if the temple has
been crushed to rubble? You're a stranger in a strange land, torn from
every place that gave your life a framework.

How do you speak to a people who have given up, who declare that
they might as well be dead—no, more than that: who say they *are* dead?
As they sat and wept by the rivers of Babylon, the people of Israel were
dead to hope, dead to a future, and worst of all, convinced that they
were dead to God.

How does a priest talk to a people when he has no temple and no sacrifice? Every tool in his priestly toolbox was gone. He had nothing. But
Ezekiel wasn't just a priest. He wasn't just a temple functionary. Having
been torn from the foundations of the past, he suddenly found himself
connecting to God in ways that he hadn't planned for and hadn't trained
for and didn't expect.

Ezekiel found God in ways that go beyond rites and sacrifices. Throughout his prophetic book, we see him moved by visions and impulses that
stem from God. Each time, it's a vision. Each time, Ezekiel is being
grasped by God's Spirit and shown something that an ordinary way of
understanding can't express. More often than not, he chooses to express

himself in odd and extraordinary ways. He digs holes in things. He lies on one side for a month. He makes bread by cooking it over a burning mound of excrement. It's supposed to be human excrement, apparently, but God relents and lets Ezekiel bake the bread over cow dung. Like I said, Ezekiel was a weird, weird dude, as much performance artist as prophet.

One of his visions told of what it meant for a people to find themselves again. The vision is one of a great valley, one that could be accurately called the "Valley of the Shadow of Death." It's filled with bones and lifelessness, the remains of a people. There's nothing there anymore, nothing at all. They are not just bones but dry bones. It's a stark and desolate image, in which no hope remains. It is the most primal form of ruin.

In the midst of this bleakness, Ezekiel finds himself set down by his Creator's hand and spirit. He surveys the death around him as his Creator shows him around. He gets asked a question: "Can these bones live?" And Ezekiel says, wisely, "Um, I think you have the answer to that" (Ezekiel 37:3). The answer rested in God's creative power, so far beyond Ezekiel's grasp that in this vision, he wisely chose not to even hazard a guess.

In this vision, that which seems totally lost—that which appears irrevocably broken—gets remade. And then it is not just remade but given life again. "Come from the four winds, O breath," Ezekiel is told to say, and he does, and what seemed broken beyond repair is made whole again (Ezekiel 37:9–10).

Dem bones, dem bones gonna walk around, James Weldon Johnson wrote in that old spiritual of hope. That was the message that the shattered people of Israel needed to hear in their lostness and despair. It is also what we need to hear when we find ourselves torn from our land, torn from our homes and in a strange new place.

In those places of forcible diaspora, what allows us to endure is patience and trust. *Patience* because as we are pushing our way out of those times of seeming hopelessness, it doesn't happen on our schedule. *Trust* that even though we may not ourselves be able to find an answer, that does not mean that there is not an answer out there. And deeper still, we learn to trust that God is with us in whatever new places we encounter. Trust can be particularly hard when we find ourselves moving not because we are seeking new opportunities but because we are being driven by forces beyond our control.

There's another vision from the writings of the prophet Ezekiel, one that speaks directly to the souls of the displaced and those who find themselves in exile. It's the vision that begins Ezekiel's prophetic ministry as he sits by a river in Babylon. He's been torn from his homeland, torn from his temple. His wife is dead. His people are without hope and in a land where they have no inheritance other than subjugation.

His entire understanding of his faith, and of his nation, was that the God of Israel rested in power at the temple in Jerusalem. Holiness was a question of place and location, of a singular holy of holies, where the priestly class maintained a containment field around the dangerous power of God. That whole way of viewing the world lay in ruins.

And then God came to him wild and mobile, winged and wheeled, with wheels upon wheels, mobile and infinitely dynamic, completely free of the false constraints of temple and ritual.

Though the messages Ezekiel shared weren't easy for those around him to hear, implicit in them was a simple truth: You may be far from home, forced to live in a land not of your choosing. But that does not mean God is done with you.

Again: even in times of desolation, forced diaspora, and loss? God is with us.

We'll need that sense of long-game, hopeful endurance. We'll need to be able to encounter God and our purpose in the new world we're soon to inhabit. Even as we fall away from the coastlines, as we lose places and cities that have come to define our sense of who we are, God will still be with us.

That sense that our self is defined by a relationship that transcends our tribal or national sense of place is necessary for two things. First, it keeps us sane and hopeful when we aren't just far from home but when home must come to mean a new thing. Second, and more vital, there is a collateral blessing. If we are aware, at a soul-deep level, that this is something that can happen to any of us, at any time, we are far more willing to be on the receiving end of others who are driven from their homes, open to the refugee and the stranger.

Or we should be, if we have even the smallest flicker of compassion in our hearts. Which you do, of course. Because of course you do. You wouldn't be a disciple of Jesus if you didn't. Let's talk a little more about that.

CHAPTER 12

WELCOME THE STRANGER

The first church I can remember from back when I was a tiny, fluttery ankle-biter was St. Andrews Presbyterian Church in Nairobi. It was the early 1970s, and my family was stationed in Kenya, where my father ran the East Africa desk of the *Voice of America*. I remember flipping through board books with pictures of a pastel Jesus and mucking about with Legos. I remember playing in the sandbox outside of the nursery, where there was a little metal bulldozer that I used to build little roads on which some matchbox cars could race. I remember my uncomfortable little outfit and having to sit still during the service. And I remember sneaking up into the pulpit while my parents did churchy things, clambering up on a chair so I could peer out over the empty pews.

Though St. Andrews was planted by Scottish missionaries, the congregation—along with most of the churches in Kenya and all across Africa—expresses Christian faith in a uniquely and vibrantly African way. When I was in my twenties, I experienced more African worship as I visited my parents at their posting in Ibadan, Nigeria. They attended a couple of churches but most frequently worshipped at the Presbyterian congregation located on the campus of the University of Ibadan. It was a short service by West African standards, meaning it went on for three to five hours. The sanctuary was filled with men and women in brightly colored traditional Yoruba garb, the air vibrating with full-throated songs and enthusiastic preaching, the room hot and rich with the sweet, mingled scents of smoke and sweat.

I can't remember ever feeling uncomfortable, out of place, or anything other than welcomed in those churches. I was something of a curiosity, of course, being an *oyibo* and all. *Oyibo* is a word used to identify a foreigner—a word that would be chanted by the crowds of children who would follow me around in the marketplace.

There's a truth to Christian faith now that's easily overlooked among American Christians: Christianity is, demographically speaking, no longer a European or American religion. It has taken deep root in Africa, South and Central America, and Asia. By the year 2060, a Pew Research Center study shows that 78 percent of the world's Christian population will be located outside of Europe and North America. Of the 3.1 billion Christians who are projected to inhabit our world in forty years, 42 percent will live in sub-Saharan Africa, 23 percent in South and Central America, and 13 percent in Asia.

Christianity is now wildly polycultural. The message of Jesus speaks in a hundred languages and in the cultural idioms of dozens of discrete cultures. It's one of the great, rich joys of the faith.

I've been to Orthodox services, with their incense and High Slavonic. I've sang and danced in urban storefronts in Black American Pentecostal churches as an exhorter belted out her percussive glossolalia. (I'm not sure if what I did could be accurately called "dancing," but it was at least an honest effort to move rhythmically.) I've heard the shouts and guitars as I've walked past the nearby Iglesia Pentecostal, the local immigrant community praising *El Señor*. The soft carillon of the nearby Methodist church still sings through the trees. I've prayed in the many-voiced prayer of Korean congregations, lifting up our hopes to *Hananim*.

The message of Jesus may have been carried out into the world on the tides of empire, as it was to my Scots-Irish ancestors when they were converted fifteen hundred years ago. But it subverts and challenges the human power that carries it, and when the powers recede away, as they always do, the gospel remains.

Where it remains, it strips away the trappings of power and takes voice in the language and culture of the peoples who come to know it as their own. It belongs to no one people, to no one place. The physical and cultural boundaries we create mean nothing at all to Jesus. The gospel, being universal, speaks its truth in every language and every cultural idiom.

As our world gets hotter and harsher, that basic reality of our faith will present Christians in the United States with a fundamental moral challenge. The most likely projections of where climate change will be the harshest just happen to overlap with the places on the globe where the message of Jesus is bearing the richest fruit. With temperatures soaring, crops failing, and desperation-fueled violence increasing, we can expect to see an exponential increase in the flow of migrants seeking respite and refuge.

And that, for Americans who claim Jesus as their Lord and Savior, is going to be a challenge. The flow of migrants and refugees on the southern borders of the United States and along the southern borders of Europe is only a taste of what will be coming as famine, drought, and war rise.

They will be Christian brothers and sisters.

It would be nice to think, wouldn't it, that a country in which a large portion of the population likes to proclaim that it is a "Christian nation" would open itself up to other Christians fleeing increasingly untenable situations in their own countries. Honestly, Christian welcome and hospitality should be open to any and all, without thought to whether they happen to share our faith. We are called not to only show kindness to those who are us, after all, but to give of ourselves abundantly to all who ask or have need.

When it comes to hospitality and welcome of the stranger, the Bible is profoundly direct. In the ancient world, hospitality toward the stranger or the one in need was a fundamental value. We, as a putatively "Christian" nation, seem to have chosen to forget this. Instead, we are doing everything we can to stem the flow of huddled masses yearning to breathe free. We create internment camps. We separate families, arguing that such punishment is necessary for people who aren't following our laws on legal

entry. At the same time, we make it harder to enter legally, cutting back
on access points, reducing the number of families and individuals we
allow in, and casting up walls around our nation.

Truth be told, there is scriptural precedent for this. In the Bible, we can
read about walls designed to keep out the stranger. Walls are, in a certain
way, unquestionably "biblical."

The first and most familiar? The walls around Jericho. Those walls
served a particular purpose: safety and security. The walls of Jericho
stood, as the story goes, to keep out a foreign, Hebrew rabble, a mass of
humanity fleeing oppression and poverty on their way to a land where
they could settle.

The choice of the residents of Jericho was *not* to show welcome. They
shut their doors against this motley army of aliens that came through
their land. On many levels, this is an understandable response. Here were
people who were taking the resources of every land they passed through.
There were so many that they seemed to pose a threat both to the culture
and to the economic structures of the region. It was not just a migration,
not just a flight, but an invasion of a host that took what it needed by force
when hospitality was not offered—which, more often than not, it wasn't.

So hiding behind a wall certainly makes sense when seen from the
perspective of Jericho.

But those walls came tumbling down, as the story goes. Those horns
must have been played by Fred Wesley. So much funk, even them walls
gonna dance.

Even so, this isn't my favorite story. Whenever a biblical tale ends with
a holy-war massacre of women and children, I cringe more than a little
bit. Because, well, Jesus.

The best-case-scenario issue here, as it is in so much of the biblical
narrative, is to see it through the lenses of hospitality. Closing doors to
the journeying stranger in need is a violation of the fundamental ethos

of every Semitic culture—and, by extension, a violation of the will of the God of Israel.

Walls appear again in the narratives of Ezra and Nehemiah, the scribe and governor charged with rebuilding Jerusalem with the blessings and resources of Cyrus of Persia. Jerusalem's new walls would permit the Jewish people to regain their integrity as a nation-state. The walls were a fresh bulwark against invasion and a mark of the renewed strength and pride of Judah. Both Ezra and Nehemiah saw the rebuilding project as part of the divine work, the will of Yahweh in reestablishing the boundaries of a shattered people.

Here it's primarily presented as a theology of reconstruction and collective identity formation. "We need to be ourselves," argue the scribe and the priest. "We have to set boundaries that keep us safe."

But there's a twist—a theological minority report, one that is woven up into Scripture. Because physical rebuilding was not the only item on Ezra and Nehemiah's agenda. They were also obsessed with racial purity, on reestablishing the blood-integrity of the Hebrew people.

So in addition to the physical walls, Ezra and Nehemiah instituted a policy of driving out the foreigner, casting out the wives and children whose heritage was mixed or impure, colored by the "lesser" peoples of ancient Palestine. To put it less gently, Ezra and Nehemiah were straight-out racists.

The Bible, being a book of many times, voices, and perspectives, resists that part of their story. That resistance comes from the book of the prophet Isaiah, which makes a fiery case for the inclusion of the foreigner and the alien. The foreigner who honors the laws and God of the land must be welcomed, saith the Lord through Isaiah. It comes from the book of Ruth, which came into its final written form after the exile. In direct opposition to Ezra and Nehemiah's racist pogroms, the story of faithful Ruth reminds the Jewish people that David, their greatest king, had her Moabite blood running in his veins.

Those walls cast up to keep the Other out, says my sacred Book of Books, do not come without peril to the souls of those that build them.

When a community chooses to slam its door on the stranger in need, God doesn't smile on that community.

And when we do violence to the stranger and treat them harshly? The biblical witness recounts what happens to people who greet the journeying stranger with violence in a familiar story: the tale of the city of Sodom.

This little story—part of the ancient histories of the Hebrew people—is conventionally interpreted in a pretty straightforward way, one that tends to be filtered through our culture's endless obsession with sexuality. But there's much more complexity to the story of Sodom than first meets the eye. The angels who visit Lot have just visited Abraham and Sarah, where they were granted generous hospitality. Being welcoming to the stranger in your midst was a core principle in the Semitic world, and when the angels arrive at the gate of Sodom, Lot insists on feeding them and putting them up.

That evening, every single man in Sodom shows up at Lot's door to rape his guests. Lot's reaction is interesting. He offers the rapacious mob his virgin daughters instead.

Why? Why is he willing to toss his girls to the crowd? In his attempts to persuade the assailants to back off, what does Lot say? He doesn't say, "Don't do anything to these men because Adam and Eve, not Adam and Steve." He says, "Don't do anything to these men, for they have come *under the protection of my roof.*"

The primary issue here, as articulated by Lot, is that the basic principles of hospitality and care for the stranger are being cast aside by those seeking to do violence to others. Violating a guest was, by the unfortunate standards of ancient Semitic culture, far more shameful than violating a female family member. Women were, after all, little more than property. Oy.

From this passage, that willingness to do violence to another for one's own gratification, rather than same-gender intercourse, seems to define Sodom's wickedness. As the text itself doesn't really provide much support for the popular interpretation of why God trashed that town, we should ask ourselves, What does the rest of the Bible have to say about "the sin of Sodom"?

In the *Tanakh*, we hear only two references that try to explain what happened in Sodom, both times in prophetic literature:

> The look on their faces testifies against them; they parade their sin like Sodom; they do not hide it. Woe to them! They have brought disaster upon themselves. Tell the righteous it will be well with them, for they will enjoy the fruit of their deeds. Woe to the wicked! Disaster is upon them! They will be paid back for what their hands have done. Youths oppress my people, women rule over them. O my people, your guides lead you astray; they turn you from the path. The LORD takes his place in court; he rises to judge the people. The LORD enters into judgment against the elders and leaders of his people: "It is you who have ruined my vineyard; the plunder from the poor is in your houses. What do you mean by crushing my people and grinding the faces of the poor?" declares the Lord, the LORD Almighty. (Isaiah 3:9–15)

> Now this was the sin of your sister Sodom: She and her daughters were arrogant, overfed and unconcerned; they did not help the poor and needy. They were haughty and did detestable things before me. Therefore I did away with them as you have seen. (Ezekiel 16:49–50)

Here, neither Isaiah nor Ezekiel uses Sodom as an example of God condemning same-sex relationships. In fact, sexual immorality doesn't come into the equation at all. If you read the plain, straight-up text itself, it just isn't there. Sodom, for *both* of these prophets, is used to symbolize those who oppress the poor, as they lay a theological whuppin' on the self-absorbed Judeans around them.

Well, what about the Gospels and Epistles? There are nine references to Sodom itself in the New Testament, and of those, eight just reference it as a sinful place. One single verse, Jude 1:7, has a different perspective than Isaiah and Ezekiel, which I'd describe here if I felt it had any bearing or if I felt Jude was the kind of book of the Bible worth teaching from. But the letter of Jude is a weird, poisonous, mean-souled missive, a little dagger of unpleasantness remarkably devoid of grace. Sure, it's in the Bible. But it's also one of those books, as Martin Luther once pointedly said about the book of Revelation, where "Christ is neither taught nor known." (It's no small irony that the same guy who gave us *sola scriptura* was perfectly willing to mock, disregard, or reject books of the Bible he didn't like. There's something to be learned from that.)

What is clear from the preponderance of evidence is that welcome is a nontrivial part of the values of the culture from which Jesus sprang. Engaging in violence against the stranger, or casting up walls of unwelcome? Those things result in the opening of a can of divine whup-ass. Biblically speaking, choosing unwelcome brings those Jericho walls tumbling down. Violating the stranger seeking hospitality in your midst brings fire down on the selfish, lustful, violent Sodomites.

Given the harsh dynamics of the ancient world, this shouldn't come as a surprise. Traveling in the ancient world wasn't easy, and finding water and food wasn't as simple as pulling off at the next exit. Wayfarers might find themselves far from a town or encounter hardship, and in that context, the ethos of hospitality was a basic necessity for the survival of a society. When you were traveling, hospitality and welcome weren't just niceties. They could save your life, which is why most Semitic cultures place such a high value on welcoming the stranger.

That emphasis on welcoming the stranger was so important, in fact, that it is written into the laws of Torah. The Jewish people are called to remember their experiences of slavery and oppression, to remember what it was like to be a mistreated alien in a foreign land, and to treat the strangers among them as they would hope to have been treated. In Torah, we hear that reminder in Exodus 22:21: "You shall not wrong nor oppress the stranger, for you were strangers in the Land of Egypt." *Surely*, you might think, *that can't mean that we're expected to treat them as if they were citizens with full rights and . . .*

And in Leviticus: "The alien who resides with you shall be to you as the citizen among you; you shall love the alien as yourself, for you were aliens in the land of Egypt: I am the Lord your God" (Leviticus 19:34).

I mean, how can these people show up and expect us to feed them and clothe them; why should we have to . . .

And in Deuteronomy: "For the Lord your God is God of gods and Lord of lords, the great God, mighty and awesome, who is not partial and takes no bribe, who executes justice for the orphan and the widow, and who loves the strangers, providing them food and clothing. You shall also love the stranger, for you were strangers in the land of Egypt" (Deuteronomy 10:17–19).

And because the Israelites were as thickheaded as the rest of humanity, the Scriptures that Jesus knew don't just leave it at those texts. In the Talmud, rabbinic tradition identifies at least *thirty-six* texts that exist to remind Jews of their duty to care for the alien, the foreigner, and the stranger in their land.

The argument could be made, of course, that these are "Old Testament" texts. They're outdated. That's simply not a good argument, given that Christianity doesn't reject, refute, or replace the heart, spirit, or G-d of Judaism. But it's an argument that gets offered up nonetheless.

In the face of that resistance, we have Jesus, who lays out with complete clarity what he expects of us in this brief, fleeting flicker of mortal life. We find his moral teachings most clearly presented in the Gospel of Matthew, chapters 5–7, which Jesus-folk know as the Sermon on the Mount. Those teachings affirm the moral core of Torah and call upon humanity to go even further and to press deeper still into the grace and mercy of God. Jesus teaches that we are to stop our endless fretting about whether we're being treated fairly and getting what's ours and instead approach our neighbors, the stranger, and even those whom we think of as our enemies with forgiveness and a generous heart (Matthew 5:17–20, 38–42).

The boundaries of our care for others, Jesus says, cannot be based on our ethnic or national identity. Our call, if we are his disciples, is to show God's love to every single person, treating them, in keeping with the highest principles of Torah, as we ourselves would be treated (Matthew 5:43–48; 7:12).

We hear all of this and sure, it's Jesus. If we say we're Christians, we're supposed to listen to Jesus and do what he says. But being as we're human and all, we tend to resist Jesus when he's inconvenient or when he makes us feel uncomfortable.

Why, we ask, should we do this? Why should we bother?

Jesus gives us the "why" in the one place in Scripture where he describes the final judgment in detail. We also find it in Matthew 25. This particular and familiar story is unique to Matthew. None of the other three Gospels contain it. It is also the only place in any of the Gospels where Jesus explicitly describes what will happen at the conclusion of all things. When Jesus speaks of the reign of God, he almost always does so using poetry, metaphor, and storytelling. These are forms of teaching that require us to use our insights and imagination and don't lend themselves to being taken literally.

Here, in the final story of a sequence of stories that have brought us to the end of Jesus's teachings in Matthew, though, he steps away from that approach. This is not a parable, a story told to speak to a meaning beyond the story. It is not an allegory in which everything in the story acts as a symbolic stand-in for some other thing. It's just a description, a recounting—or, I suppose, a *pre*counting—of how things are going to wrap up.

It's a pretty classic image, with the Son of Man on the throne of glory, surrounded by angels and suffused in the sort of light you typically see in a Steven Spielberg film. This is the big, cosmic, Sorting-Hat moment, when the lives of every single human being from every single nation are measured against the only standard that counts.

It appears, as we read it, to be something of a binary process. The good go on the right, in the sheep line. The not-good? They go in the goat line. The sheep are the righteous; the goats, the unrighteous.

Figuring out the difference should be a simple thing. All the Son of Man needs to do is check whether you're a member in good standing in the church, right? You've loudly declared Jesus Christ to be your Lord and Savior. Oh, and you've gotten your generous pledge in for the next year.

But when all those nervous Buddhists get to the front of the line, we find that the judgment call is measured by standards that are different from those we might otherwise expect. For the really amazing amount of energy Christians have spent arguing about theological and doctrinal issues over the last two thousand years, there's no doctrinal multiple-choice test administered. There are also no references, much to the befuddlement of many followers of Jesus, reporting whether you've been a churchgoing Christian or accepted Jesus as your Lord and Savior.

Instead, the measure is this: feeding the hungry, clothing the naked, and visiting those who are sick or imprisoned.

Period.

Done it? Then you're set. Somehow managed not to get around to it or found reasons not to do it? Then things aren't looking so good for you.

To be fair, there are some scholars who argue that this isn't a generic call to be welcoming and kind to all in need. I don't hold to that narrower interpretation myself, particularly in light of the teachings of Jesus on how we are to treat our enemies. But to be "fair and balanced," let's look at that particular spin for a moment.

From the narrowest possible interpretation of this text, the final judgment revolves around the term *little ones*, which in Matthew's Gospel typically refers to newly minted Christians or the members of the early church. From that perspective, what counts is how you have treated *Christians*. Did you care for them? Did you welcome them? Did you feed them, clothe them, and bind their wounds? Then you're fine, no matter who you are.

The reasons for this interpretation have to do with the context of Matthew's Gospel. Matthew is the most "Jewish" of the Gospels and seems to have been written for a community made up of people who'd been cast out of the synagogue for following Jesus. For those folks who were ex-Jews, there would have been family and friends who hadn't left with them and who'd chosen to stay in their community but still cared for and loved the Christians who'd left.

"What about those people?" asked the early church. "Are we supposed to believe that our non-Christian friends who feed us and visit us and love us are damned forever?"

So in Matthew's Gospel, that teaching of Jesus had particular relevance, and that story Jesus told was retained.

Again, this isn't my interpretation of it, but—given the likely impacts of climate change on the Christian world—it doesn't make the argument any easier for those who would advocate for limiting the flow of refugees. Given the demographics of our dynamic, global faith, who will these refugees be?

Millions of these refugees will be Christians. The same folks from Africa, South America, and Asia whose churches American Christians raise money to build, to whom we send missionary workers and Bibles, whose hospitals we fund, and whose pastors we support. When they arrive on our doorstep fleeing starvation and death, we can't refuse them and legitimately claim to be Christian.

The Christian choice, faced with the massive, climate-change-driven migrations that are to come, must be welcome and hospitality. It can't be to hide behind the walls of nationalism or to use legal structures we have created to justify driving away those who are crying out for sanctuary.

Why? Because Jesus says this is the rule by which we are measured. If you claim to be Christian, resisting his command to care for the stranger means you have traded your mortal soul for the power of the world. It also means you do not inherit eternal life. It means that no matter what words you may speak, you have chosen the goat door.

We are, of course, free to make that choice. Right now, as of this writing, it is the choice most of American Christianity has made. We've chosen power, fear, control, and self-interest over Christian morality and hospitality.

Even our prayers reflect this strange, soul-blind ethic. As I write, reports and images have been coming through, day after day, of Hurricane Dorian tearing through the Caribbean. It made a peculiarly appropriate backdrop to this writing. What was initially forecasted to be a minor tropical system, one that would dissipate after crossing the mountains of Hispaniola? It blossomed into a monster, a beast of a storm, one bearing winds and surges that meant that once again we would be hearing the words *historic* and *catastrophic*.

The forecast, twitchy and uncertain, seemed for a while to plant the storm squarely across the peninsula of Florida. Our machine-minds projected out likely scenarios, calculating and recalculating the probabilities using wildly complicated mathematical modeling. But for all of our models, the storm did what it did, growing fiercer than we'd predicted and lingering over the Bahamian paradise for more than a day, meting out destruction layered on top of destruction. We who are now used to watching storm chasers stream video and locals putting up videos were left momentarily blinded, unable to peer into winds that were more than twice the power of that derecho that tore through our area a few years back. When it finally passed, at its own monstrous leisure, it left a paradise as a ruin.

Why? Why this storm and why now? It is our human nature to wonder at the reason for things, particularly events that make us recoil in horror. In our desire for control, we want to assign blame to the suffering or to celebrate the deliverance of the righteous, particularly if we happen to have escaped this time around. But storms do what they do on their scale and not ours. God makes the rain fall on the righteous and the unrighteous alike, as Jesus reminded us in the Sermon on the Mount. Tragedy befalls both the kind and the cruel.

Against this, we wish to have power. We want ourselves, our communities, our nation to have power. To prosper. So we pray from our ego. We pray from our desire for the integrity of our material selves.

But what are my prayers against the storm? What is the purpose of such a prayer? Unlike a prayer for healing or a prayer for a change of heart, a prayer that calls for a storm to turn from us has implications beyond ourselves. Am I to proclaim with joy—as some did after Dorian—that I am certain my prayers stalled a storm elsewhere so that I and my property might be spared while others know terror and ruin instead? Am I to declare that the Creator of the Universe favors my safety over the life of a terrified child, torn from the arms of their father, water filling their lungs as they are swept into oblivion?

Let them suffer that I may not, we cry to the heavens. *Take them instead*, we cry. This is the prayer of the crucifier. I may ask for deliverance, being a human and finite creature. I will certainly give thanks for life and being when it comes. But I am not the center of things. There are times, my Master taught, when the cross will come to me, and I must take it up and bear it. It can mean the loss of everything this world offers, life included.

It seems like a thing that any disciple of Jesus would know.

What Jesus calls us to is not to live removed from tragedy. What matters is how we respond to those events that shake our lives and those events that leave the lives of others in ruin. For our own moments of brokenness and mortality, we're reminded to remain resilient, to place our hope and trust in a God who transcends us infinitely. Where we see our neighbor struggling, we're called to stand fast in compassion, helping as we can, seeing their suffering as our own, remaining all the while strong in our sympathy.

No matter how often these storms rise or how they affect us, our ethical ground as Christians remains the same. We must care for those in need or our faith is a lie. And as our world grows harsher and harder, that's the first moral ground that will be tested.

CHAPTER 13

SLOW DOWN, FIND SABBATH

The night was impossibly gorgeous, as a day that seemed more like one in late April than late July wafted cool and lovely over the stress-shimmer panic of the endlessly moving Washington, DC, bustle.

After working on a novel manuscript for a while in the library, I stepped outside to take a long walk. It was a little after eight o'clock on the kind of evening that makes you think of Yahweh walking bare-toed and joyful in the garden. It was the veritable Platonic form of the cool of the day. The sky was lit with clouds still catching the last rays of the setting sun, set salmon against the clear, darkening blue of the sky, and the air was crisping up nicely.

Perfect. I love long walks on perfect nights that clear my mind and wake my soul.

I walked through the neighborhoods of Vienna, Virginia, which, when I first moved to DC, was an outer suburb. As part of that first tide of tract development, Vienna filled with humble brick ramblers and ranch homes, which quickly filled with midlevel federal workers and army officers.

But the tide of development has swept on, and Vienna is now part of the inner ring of suburbs. Better yet, it's served by a Metrorail station. That makes it highly desirable for those humans who prefer not to spend two to three hours of their waking day stuck in the endless columns of traffic leading to the exurbs. As Dante wrote, the worst thing about the deepest layers of hell is that you have to commute through all the others to get there. That was the point of the chapter in Dante's *Inferno* on commuting, if I'm remembering correctly, although it's been a few years since I read it.

Now that Vienna is a place where people want to live, it has begun to change. Walking through the neighborhoods shows a sea change in the homes. Those old tract ramblers, identical to the one I inhabit? They're being snapped up, razed to the ground, and replaced with much larger homes.

I'd call them McMansions, but that would be inaccurate. They're as *big* as McMansions, but the people buying into Vienna are folks of wealth and taste. No giant, ticky-tacky slabs for them. These are homes for people who know the meaning of the word *bespoke*. Almost without exception, they are lovely. They are big and gracious, with elegant columns and careful detail work. They evoke the nicest house in the nicest neighborhood of a large Southern college town—the one the president of the bank would have lived in, only updated.

And because they harken to that bygone era, they all have porches. Some are wraparound. Some are front porches. All are deep enough for a small party, and most seem to be furnished with gracious, pillowed wicker offerings from some fancy bespoke outdoor furniture store.

What struck me as I walked past home after beautiful home were two things. Although it was after eight thirty at night and dark was falling, most homes appeared empty. The windows were dark. There were only sporadic signs of life—a walker here, a jogger there. But most people were either still working or else kid shuttling, rushing their young ones from one activity to the next.

My own son, after all, was in just such an activity, which is why I had driven thirty-five minutes through the worst rush-hour traffic in America to get to Vienna in the first place.

Occasionally, a late-model Mercedes would pull into a driveway and slink elegantly back to a tasteful garage. In one lit window, a man in his late fifties sat in a beautiful home office, talking intently on his iPhone with the huge screen of his iMac open to email.

What hit me harder still was that of the fifty porches I passed—gorgeous, elegantly decorated porches, on the most beautiful evening we'd had in months—not a single one was occupied. If ever there was a night for a porch, this was it, appearing like a miracle in what should be the soup and swamp humidity of a DC July.

But they were all empty. No one was gazing out at the fireflies. No one was rocking slowly in a wicker chair and nursing a cocktail, shooting the breeze with a friend. There were no clusters of children scampering and squealing and delighting in the free fellowship of friends.

At this, the prophet Isaiah whispered in my ear, as he so often does, "Many houses shall be desolate, large and beautiful houses, without inhabitant" (Isaiah 5:9).

So much of American life now is strangely anxious, leaping and rushing and impatiently chasing from one instant to another. We frantically juggle the two jobs we now apparently need to survive. Fearful for our children's future, we hover and manage their activities, carefully entering the time allotted for their friendships into our shared family calendar app.

We "helicopter."

When my children were young, I was not, or at least I tried not to be, a helicopter parent. You know the sort: the ones who schedule every last moment of their child's lives and whose rotor wash manages to blow every last particle of faerie-dust whimsy out of childhood.

I did not wish to be that sort of parent to my boys as they grew up because that approach to children has very little to do with loving them and a whole bunch more to do with our own anxieties about ourselves and our world. It can, as a stream of fretful articles highlight, cripple the development of our children. So protected, they become vulnerable; so carefully managed, they have no idea how to live for themselves.

Ahhh! We're so anxious, we're anxious that we're anxious! We're meta-anxious!

That anxiety is everywhere in DC. I see it as I walk. Walking, again, is so much better than driving. It allows me to go slow and take time to really observe the world around me.

On a similar summer night in that same neighborhood, I walked past one parent who was sitting outside of his child's music lesson. His car was idling with the windows up on a beautiful, late-spring afternoon, and he sat inside, fiercely and loudly arguing with his spouse about their schedules on a cell phone. He was oblivious to my passing, my hearing the details of his anger and his frustration. There was no world outside of the cell of his car and the cage of his calendar of events.

I encountered another adult just a few minutes later. She was the loudest of a cluster of parents shouting instructions on a sports field, running the carefully scheduled activity that now fills the time that once would have been filled with childhood's freedom.

"Watch me, all of you!" she barked on the softball field to a gathering of ten-year-old girls, all helmeted and wearing complex black metal face guards. Face guards? Since when did softball require a mask for every single player? There's a small fortune in orthodontia to protect, I suppose.

"This is how you call it," she shouted, motioning to one of the five other parents to knock a ball skyward. "MINE MINE MINE MINE!" And she caught it cleanly.

"Again! Watch me, Tyler!" Tyler and the other girls seemed distracted because there was a bell.

Music—an ice cream truck chiming its seductive music across the field.

"Tyler! Eyes on me! *Now!*" Another ball was summoned, and the parent cried, "MINE MINE MINE MINE MINE!"

The ice cream truck rang its bell forlornly in the parking lot, but there were no takers. Though the park was full of children after a long day of school on a warm, early-summer evening, these were not children at play. They were on task.

That task is the rush of activity parenting, and I will honestly admit that I wonder at our hubris in the context of this crisis. Here are moms and dads who genuinely love their kids and hope for their future, bundling their beloved offspring into their SUVs to drive for hours to softball and soccer and baseball, to tutoring and enrichment activities. When an entire culture of hundreds of millions organizes around anxious hyperactivity as a social expectation, that has an impact on how badly this crisis will affect the lives of those same children.

How many gigatons of carbon have been released into our thin sliver of atmosphere in the service of getting kids to travel-league games and practices? How much have we sacrificed the future of our children's children on the altar of swim meets? It seems so . . . absurd. It's just not a good trade at all.

I've talked to my sons, now grown, about the pace of their lives growing up compared to the pace of my life just a generation before. As much as I resisted it, we still did succumb, enough that there was a certain generational jealousy for the pace of my childhood and adolescence. I wasn't rushed. I had time of my own. This is, I think, one of the reasons for the peculiar resurgence of reminiscences about the 1980s in the last few years.

I don't remember the Reagan years with much fondness. There was that threat of nuclear holocaust pretty much at any moment. It was the "greed is good" era, when America buried itself in the false glitter of debt. It was the "War on Drugs" era, when millions of Americans, disproportionately Black, were imprisoned for nonviolent crimes.

But it was much slower. The pace was more human. If you look at films from that era, particularly the works of Spielbergian Americana, you see a very different culture. Look at the neighborhoods in *E.T.*, filled with packs of kids just doing their thing. How many times has your kid gone out with a pack of neighbors on bikes, returning only at dusk, dirty and tired and harboring an alien? Or if your progeny are as introverted and bookish as I was as a lad, how often have they come home, done a mediocre but acceptable job of completing their homework, and then lost themselves in a novel for the rest of the evening?

We Americans just don't do that slower, human-scale life anymore.

And it's not just making us anxious. As we do it by the hundreds of millions, it has changed our planet.

Our mad rushing about isn't just limited to how we raise our children in this generation. It's also how we live our lives. Work, for us, never ends. It's on, 24-7. Our laptops, always open. One more email is always there, demanding a response. We are flying from place to place, having important meetings to discuss the next important meeting we must have. We are in our cars, still connected, taking calls while driving on our way to a meeting. We have, it seems, never been busier.

But busier doing what? What are we actually doing? Is it important? I mean, really and actually, does the thing we spend our lives racing about doing and stressing about make a difference?

Americans just can't seem to live at a sane pace.

Every few years, there's another spasm of interest in this paradoxical characteristic of the American economy. We work, and we work a whole bunch. In major metropolitan areas like my own, the number of hours you put in each week is almost a badge of pride, yet another way in which you can prove yourself superior to those around you. Downtown, little clusters of young associates regale each other with tales of epic hours worked—fifty hours, seventy hours, eighty hours, one hundred hours in a week. Among some junior partners at area law firms, where racking up billable hours is next to godliness, there's a rumor going around that if you pound back twenty-five triple espressos in a row, it actually rips a hole in the space-time continuum, allowing you to put in that perfect two-hundred-hour week. Either that or your head implodes—and honestly, after twenty-five espressos, it's a little hard to tell the difference.

This ethic of intensity expresses itself, oddly, in the work schedules of many of my pastoral brethren and sistren. I'll talk to minister colleagues who'll tell me they don't have ten seconds to rub together, that they're putting in two full weeks every week, that they're exhausted and overburdened. They talk a great deal about self-care, about how hard it is to carve out time for life and friends and family. This confuses me because for some reason, I have never equated spiritual leadership with shimmering stress and fatigue. But pastors, we who so often need to feel needed and want to feel wanted, imagining ourselves to be the fulcrum upon which the entire future of Christ's kingdom relies? We're just terrible at that.

Pastoral stress seems to get most intense during Lent and Holy Week, and that's the time I feel almost embarrassed every year, out of sync, and like I'm doing something wrong.

Every year right around Holy Week, at the very height of the liturgical season, I'm not even faintly stressed out at the prospect. Oh, sure, it's a little busier. I'm at church a little more. But my little church just does things simply. We have a humble meal on Maundy Thursday, one where we sing and pray, break bread, and share the cup just like Jesus asked. We fold that into a potluck meal. But it's not particularly complicated. I pick up some loaves of tasty challah and some Trader Joe's soup. We sing "Let Us Break Bread Together" a cappella, one verse before the meal, another after. That's about it. It feels holy.

Then there's a wonderful lay-led event for Good Friday as the story of the passion is retold. I show up, and I worship, and I let others lead because that's just what we do.

On Easter, I get up at four o'clock in the morning so I can get to the top of Sugarloaf Mountain in time to join with other pastors to lead the multichurch sunrise service. That's early, sure, but that's why the Good Lord made coffee plants. It's beautiful, valuable, and significant. I show up, and I offer prayers and readings and help lead the gathered folks from all around our community in song.

And then there's a brunch, and kids scamper around. Then we worship, and it's awesome.

So it's busier. But it's entirely manageable. I look forward to it theologically, personally, and spiritually. It sings to and sustains my soul.

That is not what I hear from my colleagues. It's red-alert panic mode for many, as sixty- to seventy-hour church weeks and overpacked suburban family lives collide with a blinding flurry of additional Holy Week demands. It's the liturgical perfect storm, a great terrifying wave of intensity layered upon intensity.

When I hear the war stories, I'm abashed. I can't contribute. I feel I can't even speak: "Oh, yeah, I'm totally chill. Looking forward to it! Not a big deal."

I feel . . . invalidating. A little subversive. Perhaps a little annoying.

But then again, perhaps it's worth saying. Because as much as work stress might feel like you're being flogged, being crucified, and dying—a kind of organizational reenactment of the passion—I'm reasonably certain that's not a healthy spiritual place for pastors. Or for beloved communities.

For our souls, and for the well-being of our world, we need to slow down. We Americans in particular. According to data from the Organization for Economic Cooperation and Development, Americans work more than most of the industrialized world. In 2019, the average worker in the United States put in 1,779 hours a year. That's now almost two full weeks more than the average Canadian and 350 hours a year more than the average Dutch worker.

For all of our busyness, though, we're still nowhere near matching the industrialized country that holds the record for the most hours spent working. Is it Japan? Korea, perhaps? Nope. Germany? Hah. The Germans work only 1,386 hours a year. They put in ten full workweeks less per year than we do. The people who work the most live in Mexico. At 2,137 hours annually, Mexicans, according to the data, are the hardest-working people in the world.

What's strange about this whole phenomenon is that it's exactly the opposite of what people used to think the future would look like. Back in America in the 1950s, everyone was absolutely convinced that sixty years in the future, we'd all be working fifteen hours a week. (They also thought we'd be commuting via jet pack, and I don't know which failure is more disappointing.) Strangely, though, studies have shown that when you take into account all the increases in technology and productivity, it should only take a modern worker eleven hours to do the work that took

forty hours to do in 1950. If we were willing to accept the same s̶
of living as 1950s Americans, an eleven-hour workweek would be e̶ ̶.̶.̶ely
possible.

But we don't want to live in little 1950s houses. We want to live in huge houses. We don't want to own just one car. We want three cars, which we'd put in a garage that's bigger than that little 1950s house. Heck, we could live in the garage, if the garage wasn't completely filled with stuff. We don't want just one 9-inch television. We want a 108-inch LCD HDTV so we can see the oil glistening in the pores on Jack Bauer's nose. We want our cable, and we want our fiber optics, and we want our high-speed internet, and we want our smartphones.

We *need* these things if we're going to be happy. Because we are so much happier now than we were half a century ago. Aren't we? We work so hard, every day, to gather in the material blessings of our Tantalus consumer culture, sure that our joy is just that one last gadget away.

We find security in those things, in the having and the holding, and we allow ourselves to imagine that they matter.

Though the story is now a couple thousand years old, it is that same ethic that Jesus is addressing in a story we hear told in the Gospel of Luke (Luke 12:13–21). Jesus is standing before a crowd that has gathered to hear him share stories and riddles about the nature of the kingdom of God. He's showing them what's important and telling them what they should value in their lives. He's just told them to trust the Holy Spirit to guide them in what they have to say when someone from the throng pipes up.

"Teacher, tell my brother to divide the family inheritance with me" (Luke 12:13). The audience member was hoping to get Jesus to act as most rabbis would have acted, which is to go into a long discussion of the laws of inheritance and to come up with a legal ruling for him.

Jesus turns him down with surprising gentleness and instead uses the request to launch into a story about a man who has so much stuff that he is thinking about building a new five-car garage with a finished attic for storage. This is someone who has succeeded by every single standard of classical wisdom. He has invested wisely, he's planted the right crops, and he is doing absolutely everything right. The wealthy man smiles to himself, sure that he's going to have a chance to kick back and enjoy the bounty he's gotten.

But as Christ tells the story, that's not what happens. It's at that moment that God appears to the man and berates him: "You fool! This very night your life is being demanded of you." In an echo of Ecclesiastes, we hear, "And the things you have prepared, whose will they be?" (Luke 12:20).

This is the point in countless TV sermons where your friendly neighborhood televangelist would start talking about not storing up treasures for yourself but being rich toward God—something you should definitely keep in mind when you see the toll-free number on your screen asking you for your credit card number. But that totally misses the point of what Jesus is talking about.

Within the story, what is being demanded of the rich man isn't his barns or his crops or his goods. What's being claimed is his life: all of his days, all of his actions, all of the choices he has made. The Creator of the Universe couldn't care less about possessions. It is life that God demands.

That's one of the primary challenges facing us in our modern culture of work. A deep, personal commitment to excellence in all that you do in the working world was viewed by the Protestant reformers as a sign of spiritual maturity. God has indeed given us all certain gifts and called each of us to a particular task in life, and our willingness to embrace that task and pursue it joyously is a sign of blessing.

But we're not called to pursue work for the sake of profit alone. We're called to work because work-as-calling is a joyous and honorable thing. We're each given a vocation as a part of contributing to the broader good of God's creation.

What work should *not* be, if it is to be part of the broader good, is all-consuming. If it devours time for friendships and fellowship and takes away those moments that should be given over to prayer and joy and thanksgiving, then it has grown beyond its rightful bounds.

In the face of how our anxious overconsumption and overproduction is stirring up our little corner of creation against us, we are, again, given a moral choice. We can continue to put pressure on ourselves, our children, and our world to produce and consume things that serve no purpose beyond momentary profit. We can continue to chase after the wind, our hands filled with toil but no peace in our minds (Ecclesiastes 4:6).

Or we can live differently.

The word for that different way of life is Sabbath. I know, from the great story of my faith, that all of creation demands Sabbath time. We need it for our sanity, but it goes deeper than that. Sabbath is the practice of not milking every last moment, of not squeezing the very last drop of profit from life. Sabbath is the antithesis of the order imposed by mammon.

As members of a culture that idolizes striving and structures life around endless grasping, we likely look at the idea of taking time for intentional rest as tearing at order itself. From the vantage point of our overscheduled, jam-packed lives, we encounter Sabbath fearfully as that place where we might permit chaos and disruption to enter. Sabbath, after all, is that place where we stop pouring ourselves into the structures that frame our lives, the schedules and demands and expectations that leave us continually anxious, ever behind, always stressed and struggling. Sabbath, we might think, is inchoate nothingness, not reassuring form. It is anarchy, not order.

So we impose order, layering structure on structure until there is no space for anything that is beyond our own control.

This is why we fail.

A Sabbath day is not random, and it is not disordered. It is free, yes. But when I take Sabbath times—long meditative walks, time to write or draw, time to read freely and grow in connection to another soul's writing, time to spend with friends and loved ones—they do not feel like chaos. They aren't chaotic at all.

They feel calm. Sabbath feels serene and ordered and at peace. Sabbath is not "doing nothing," at least not in the sense that we interpret it from our busyness. It's a different order, a different way to live life.

The structure of Sabbath matches the intent of creation. Our crazy, competitive, acquisitive "stresstival" of striving does not. That way of life is a poor match for the Creator's intent. It is the berserk fretfulness of a grasping culture that feels most like chaos. It is a structure so poorly suited to change that it shakes us violently, tossing us about as it itself is tossed.

What does striving look like? What does the order of Sabbath look like?

As I visualize this, I find myself seeing an image from a movie, as I often do. It's a scene from the movie *Contact*, a delightful bit of hard sci-fi directly from the mind of Carl Sagan. In the scene, Jodie Foster is being sent to meet with extraterrestrial intelligences that have contacted humanity in a device whose design has been provided by those intelligences. It's a perfect sphere, one meant to travel impossible distances across space. Humanity has made one modification to the perfect design provided to us. Inside the sphere, there's nothing—nowhere to sit. Just empty space. Surely that's a flaw, think the engineers; a space traveler would just bang around in there.

And so they build a massive command chair. It's bolted in by large, metal support columns, and as Ms. Foster prepares to go meet the aliens, she's strapped in. For control. For safety.

That chair, my metaphor-mind suggests, is our scheduled life. And Sabbath? Sabbath arrives midway through the scene. It's been a wild phantasmagorical ride as the craft dives into one wormhole after another. The command chair is shaking violently, the journey through space getting rougher and rougher. Entering another wormhole, Ms. Foster is being thrashed about in that seat like a rat in a terrier's jaws when she notices something.

She sees a watch floating serenely next to her in zero gravity. It's not shaking. It's not about to be torn apart. It's just fine. You see the light suddenly dawn on her as she realizes what is happening.

She unbuckles herself and finds that, after she is released from the vio-lently shaking command chair, the ride is actually smooth. Calm.

As designed.

Living into Sabbath is like that. Living a life that is intentionally slower and more gracious is like that. For followers of Jesus in a time of climate crisis, that's one essential part of the moral teaching of our tradition that we'll need to recover. On a world that is less resource rich and where the cheap and easy energy of fossil fuels is fraught with existential peril, we'll need to calm down a bit.

In some ways, all we have to do is *less*.

There are other ways we can do less, slow down, and throttle back on our spendthrift consumption of resources. I take my motorcycle to a nearby shop for any repairs I can't handle myself. It's a funky little local shop located between a tobacco paraphernalia store and an adult enter-tainment emporium, a far cry from the shiny megadealer nearby. The mechanics there are cheery, heavily tattooed, casually profane, and they can do the work in good time and for about half the price.

The challenge is getting back home after I've left them the bike and then getting there again. The shop is about four miles from my house.

So when I go there, I take the Metrobus. Here in this country, We Hate Buses. Americans are carefully programmed to hate taking the bus, a process that began for me in high school. It was cemented when that senior with his jacked-up, fire-engine-red '71 Chevelle reminded sad little sophomores that they weren't ever going to get a girlfriend if they didn't get themselves a sweet ride first. We don't like the bus. The popular, suc-cessful kids don't ride the bus. And we have to sit there, waiting, waiting, waiting, never sure when it will arrive. It's frustrating, particularly for thems of us who want it right now. It's the American way, baby.

That waiting impatiently part is no longer true. In DC, the buses are all outfitted with GPS transponders. Those transponders relay data about their real-time locations to a Metro computer, which then relays data to the Metro website. Information about when the bus will show up at any given stop can be accessed via your mobile phone. It's accurate to within thirty seconds, meaning you don't have to wait around for a half hour at the bus stop. It changes the whole equation.

It's not instant. You can't just up and do it. And you are standing there for a bit, looking at your phone, looking down the street. You feel the sun on your face. If it's cold, you'll taste the cold in the air. There are moments of downtime, where you're thinking or just watching the world go by. Still, it's more than manageable. You can work around it.

I rode the bus twice the last time my bike was in the shop. Once the bus was half full. Once it was nearly completely full. Both times, I was the only white person on the bus. Everyone else was Latino, or African American, or Asian, or of Middle Eastern descent. None had about them the trappings of affluence. There were no suits or expensive shoes. These were folks who are not on the upper or even middle rungs of the economic ladder. They were working class, on their way back from getting groceries.

The upper and middle classes were in their cars, zooming around us and talking on their cell phones. Most were driving SUVs. A few were driving hybrids, and fewer still, electric cars. What struck me, sitting there on my honkey behind, was just how much better at caring for creation my fellow bus riders are than I am.

Our bus was powered by natural gas, meaning far fewer emissions and no reliance on oil from nations that hate us. One-third full of passengers, the bus gets the same mileage per passenger as my motorcycle. With a full load, that bus gets the equivalent of 165 miles per gallon, which even the most efficient shiny new $55,000 Tesla can't offer.

As I looked around at America's bus riders, I realized something. For all of the talk on the right about energy independence and all of the talk on the left about environmental stewardship and carbon neutrality, the people who are doing the most to make us more efficient are the folks who *need* to be efficient.

As human beings, those who aren't growing temporarily rich on the fleeting abundance of our carbon economy are the ones who are doing the most to care for creation. Neither the folks who cried, "Drill, baby, drill!" nor the folks who buy $32,000 hybrids loaded with electronic gim-crackery (that would be me) are even coming *close* to the folks who ride the bus day in and day out.

Yeah, they may not be the cool kids. Or the "successful" ones. The journey on the bus takes more time. It isn't as immediately productive. But those things don't really matter now, do they? "Better the poor whose walk is blameless than the rich whose ways are perverse" (Proverbs 28:6 NIV).

An environmental ethic grounded in the practical, immediate needs of a community can easily rise from the faith and life of a congregation if it's led with vision. Years ago, when I was in seminary getting my doctorate in churchy churchiness, I got to know a young pastor with just such a vision. Rev. Dr. Heber Brown III is the pastor of Pleasant Hope Baptist Church in Baltimore, and his sharp mind and warm spirit lit up the conversations in my doctoral cohort.

Heber's congregation in Baltimore had recently converted their church grounds into a community garden—and not just for the simple pleasure of gardening. Baltimore neighborhoods, like so many Black communities in America, were radically deprived of access to fresh, healthy, locally grown produce. This deprivation hit particularly hard in the unrest following the funeral of Freddie Gray. With a curfew in place and most stores shuttered, food was hard to find. People knew the church was addressing that need and reached out.

"Because food was our calling card, the phone started ringing," Heber said. "After two and a half weeks, I realized we had the beginnings of a system, and it would be advantageous to keep that going." From that place of basic need, a nonprofit was formed. The Black Church Food Security Network leveraged the experience of Pleasant Hope into a replicable model for congregations, showing the path to create gardens and markets on church land.

We are in a time that will challenge us locally and challenge all of our localities together. Getting to a place where we are living a more balanced, measured existence will be harder than we think. It's easier to be anxious, to be in constant motion, to try to control our world with our overwork and our rushing about. Catch your breath. Slow down.

Remembering the best graces of a slower life—patience, neighborliness, thrift, and a sense of connection—will be one of the traditional Christian virtues necessary if we are to thrive in the world that is coming.

CHAPTER 14

LIVE HUMBLY

So we need to slow down. But we also need to consume less. We need to live leaner and live humbly.

For the last nine years, I've served a little Presbyterian congregation in Poolesville, Maryland. It's sweet and it's small, and like most healthy, small congregations, it's aware of its own place in the world. It's a kind-hearted little Jesus tribe sitting on a plot of land near a crossroads in town. We're a peculiar mix of progressive and conservative souls who've generally committed to getting along with and respecting one another as brothers and sisters in Christ. We've got three buildings on our sprawling campus. One is a simple brick sanctuary that starts feeling full at around forty souls. It was built in the mid-nineteenth century. It is comfortable, cozy, and the farthest thing from fancy. Another building is our "manse": a house that contains the pastor's office, a couple of classrooms, and a parlor for small meetings. It was built in 1827 and is, on good days, mostly waterproof. We've also got a simple 1970s-era meeting hall, one that was built by church members. It's got a kitchen, a utility space, a classroom, and a playroom.

Oh, and bathrooms, because there are no bathrooms in the sanctuary.

It's a humble place, and we have humble aspirations. We're not growing into a megachurch because there are plenty of those already. It's also not the spirit of the church, which has been an intimate fellowship for more than a century and a half. We're content supporting local missions, inviting in strangers, housing the homeless wanderer on occasion, and witnessing our community by living out the gospel. We pray. We support one another. We worship and study. Where opportunities for gracious connection with others arise, we take them.

Our property hosts the community garden for the town, in which the church invested time and our limited treasure. Even though we're in an agricultural area, not everyone has access to land of their own on which to plant, grow, and harvest. During the summer, little clusters of towns-folk pull up into our parking lot, carrying with them trowels and baskets. They return with their baskets loaded with produce. In our simple way, we've given our neighbors welcome, rich soil and water. We've shared our land and been hospitable. We do what we can do.

Ours is a pretty simple way of being together. That's not to say that life in the small church is without challenges. It's hard to be small in a culture of striving for bigness. There are times when things are tight. There are places of challenge. In the face of those challenges, we work to live within our means. We don't debt-fund growth but instead do what we can with what we have. We keep a financial reserve for times of crisis because you just never know when a pipe will burst, an air-conditioning unit will fail, or a global pandemic will put the kibosh on in-person services for nearly a year.

When the church rainy-day fund fills up and spills over, we make improvements and upgrades. But those upgrades aren't about getting bigger; they're about keeping things from falling apart and making things better for the future. Even though our facility requires the sort of maintenance you'd expect from a building that is more than 150 years old, we take the time to reduce our energy footprint.

For example, we replaced one of our aging, ailing HVAC units. Our sanctuary is now more comfortable and uses less energy to heat and cool. We upgraded our insulation, again reducing heating and cooling costs. We replaced the old, leaky, 1970s aluminum windows in our multi-use space with new, efficient windows. We looked into going solar, as visions of those panels danced like sugarplums in our heads. But we soon realized it was beyond our means as a community . . . so we replaced our old, oversized water heaters instead to cut our electricity usage.

That approach does two things. We keep our building maintained, sure. But because our maintenance is guided with prudence and directed toward being more efficient, we're also spending less to support our basic operations. Which means we're able to more easily replenish our reserves. Which means we have more to use for our mission in the community. And so on and so on. It's a self-perpetuating upward spiral.

One of the things that frustrates me most about our national jabbering about energy use is the lack of emphasis on this rather simple virtue. Listening to folks who claim to be conservative snipe at efforts to conserve and encourage thrift as somehow bad for America is just so wildly ironic. Since when did profligacy, debt, and self-indulgence become conservative virtues? None of the conservative souls I personally know and love lives like that.

It also shows a fundamental misunderstanding about the greatness of our nation. At our best, we Americans are practical people. We like to solve problems and get things done. We also like to see progress, to see those improvements bear fruit. As we try to figure a way out of our dangerous addiction to fossil fuels, encouraging this approach to life seems the rather obvious way out.

And yet we fight it, having been taught to be dutiful little consumers. Lean means we are inferior—that there is less of us, less of our possessions, less on our plates, less power. Less, you say? Less is bad! Smaller is worse! How can we possibly survive on less?

To which I'd ask most of us, When was the last time you cleaned out your fridge?

No matter what the time, or the season, there seems to be a single truth in American family life: we really, really need to clean out the fridge.

It starts in a well-intentioned way. You need to have food, after all, and there the fridge is, this giant behemoth of a thing, towering in your kitchen like the monolith from *2001: A Space Odyssey*, assuming Kubrick's monolith had been stainless steel and had a water and ice dispenser. Our fridge is, like all fridges these days, a colossal thing, a giant yawning chasm.

Back in the day, 9.5 or 10 cubic feet was all the space a household needed to store food. Now, though, we need more, because of course we do. Our fridge is huge. Like the world's largest and coolest Russian nesting doll, our fridge has enough storage space inside it to hold the fridge that cooled the food in my grandparents' kitchen, which in turn could hold the dorm fridge I used for three years of my life.

We have a huge fridge, which comes with the temptation to fill it with food. We wouldn't want it to be empty now, would we? It would be so sad. Poor, sad fridge. So we fill it, and it contains so much more than we need and, worse still, more food than we can *remember*. The food we bought a week ago gets pushed to the back of the fridge, where it's out of sight and out of mind. The food in the giant pull-out freezer? It piles up, new items on old, layer upon layer. What lies at the very bottom of our freezer now? I don't even know, but if that label-less packet of leathery, freezer-burned mystery meat turned out to be mammoth chunks, I wouldn't be surprised.

And so, every rare once in a while, we'll clear that freezer out. It's not a pretty thing because I hate throwing away food. I hate it because it's a waste, and I hate it because we do a tremendous amount of throwing away food in the United States. It's an amazing thing, actually, just how much food we manage to not use in America. This land is still amazingly fertile. The amber waves of grain that sit between our purple mountain majesties upon our fruited plain produce so much food that it boggles the human mind. Four hundred and thirty billion pounds of food a year, when you look at the numbers, which is . . . um . . . rather a lot.

Of that, we don't use 31 percent. That's right: 31 percent goes to waste. This is food that gets discarded because it doesn't meet certain specifications for appearance or goes bad on store shelves. This is food that we buy but stash away until after it's gone bad.

That's 133 billion pounds of food every single year. A recent study calculated that out of the 141 trillion calories in that food, the equivalent of 1,200 calories per American per day is going to waste. Enough goes to waste in America, in other words, to feed three hundred million people breakfast, lunch, and dinner every single day.

It's a little crazy, but it's even crazier that we don't notice. But we don't, because most of us are not hungry.

It's easy to miss things when you're living in a comfortable, abundant world. That's why the prophet Isaiah preached as he did to the overfed, complacent folks he encountered in Jerusalem. In Isaiah 25, we hear a story of new things. It's a lovely little story about a meal and about the end of suffering. It's the promise that tears will be wiped away and that God will make everything all right for the people of Israel. What we have not heard is the chapter that came right before it, because that one is a bit harder on our ears.

Chapter 25 of Isaiah comes across nice and easy, but it's part of that eighth-century prophet's many oracles against the people of Israel. Speaking from the comfort of royal Jerusalem, where he was well regarded by all and had the willing ear of the king, Isaiah could have just told folks what they could see of the world immediately around them. Jerusalem, after all, was a place that was prospering. It was the center of power in Judah, and in every age and every time, power draws wealth to itself. The city was doing well, was comfortable, was at ease. There were other prophets, I'm sure, who proclaimed to the people at that time that all was well, that everything was going swimmingly, and that all anyone needed to do was just keep on keeping on.

But that was not the message Isaiah bore. In the verses before we roll into chapter 25, we hear God's annoyance with the selfish indulgence of the world. And to those Jerusalemites, we hear a clear message: this is not a feast just intended for those who prosper now. God's care extends particularly and pointedly to the poor and the struggling. And to the denizens of Judah's capital, Isaiah says, You can't ignore this.

Following this comes a psalm, that song of praise that fills the twenty-fifth chapter. In this, Isaiah proclaims that with the collapse of the life we had known, something truly new and more gracious and more promising will arise. All will not forever be wreck and ruin.

At the heart of that vision lies, as is so often the case with Isaiah, a meal. What he brings to the table is the vision of a feast, a table brimming over with good things. It's not just intended for his few chosen or for those who are wealthy and powerful. It's intended for all peoples and all nations.

That, as Isaiah speaks into the ear of the powerful denizens of Jerusalem, is the vision of God's presence on earth, the work that God engages in as the reign of God is shaped.

And that, insofar as Jesus-folk claim to be moved and engaged by the Spirit of God at work in us, is kinda, sorta our job too. Let me say that again: if we want to claim that God is at work in us, we have to be working toward God's goals in this world.

Sure, we often encounter God as abundance. But our task, as we engaged with the great, groaning table of our formerly sweet little planet's productivity, was not to devour all for ourselves or to be oblivious to our impact. It was to be aware: aware of how our actions and our choices echo out across creation and how those choices impact the world around us as they pile up by the thousands and the tens of thousands.

Here we were, placed in an Eden that had all that we might need, that created and produced everything that we human beings required for life and then some. And we squandered it, wasted it, abused it, and took it for granted.

That is a luxury that is coming to an end. There is no reason to believe that our world will be able to produce agriculturally at the same level as the temperatures rise. What can we do? It all has to do with our attention. We have to pay attention to how we use what we have in our homes, in our communities, and in our culture.

We can attend to ourselves—to the way we approach the food that pours out of our culture's cornucopia in such a relentless flood of calories.

So much of justice comes from just paying attention. So much of justice comes from moving and acting in ways that reflect our attentiveness not just to our own lives but to the impacts that life has on others.

We can make a point of growing our own food.

Now, you'd think that'd be exactly the opposite thing we should do. Wait, don't we have waaaay too much already? There is plenty, vastly more than we need. This is true.

But the primary issue here is inattention. It's losing sight of the reality that creates what we consume. If we don't see waste as waste, don't make the connection, we're not going to notice it in our own actions. We just discard thoughtlessly, without noticing.

When you've tilled the soil, worked the earth to bring up those tomatoes and beans, that kale and zucchini, you feel it when the fruit of your labors goes to waste. And that increased awareness means you feel it more when you see other food go to waste.

And we'll feel the loss on a new and harsher world, where crop yields will be diminished. With possible agricultural production losses of between 10 and 20 percent by the end of this century, our profligacy will no longer be sustainable.

We need to be less wasteful and more aware of the preciousness of the fruit of our world. We need to be more aware of the need that will come as our new world yields less to meet our appetites. The more we connect with our world, the less we seal ourselves away from creation in the bright, hard cells of our busyness. We need that sense of connection, that deep engagement with the processes of our changing ecology.

All summer long, I tend to my beans. They aren't much to look at, those little bushes, filling a three-by-ten-foot plot to the right of our driveway. In those first days of summer, they have a hard start of it, as life often does. Some never make it much past being shoots and then wither and die. Others look promising at first, but then marauding bunnies and hungry deer nibble off their tender first leaves, and starved of life-giving light, they harden into dead twigs.

Every morning, I tend them. A little weeding and a little water as needed. Where one dies or is devoured, I replant. Some struggle to stay upright as they grow, so I place a little stake next to them and give them a boost toward the sun.

By late summer, they still aren't much to look at. Just a dozen or so nondescript plants, the tallest barely reaching my knees. But under their leaves can be found, every other day, a festival of green beans. They spring into being, leaping forth from the tiny white flowers.

I still tend them and water them as needed. And every other day, I spend ten minutes harvesting. For a couple of months, we have beans enough to fill our plates twice a week. More than enough, in fact. Bags full of freshly picked beans often make their way into the hands of family and neighbors. Usually, they're welcome.

As the growing season draws to a close, though, the time comes when those humble little bushes stop yielding. And having spent the summer with them, I feel their identity as living things rather more intensely than I would had I bought a plastic bag of flash-frozen beans from Trader Joe's.

They're pretty basic beings, green beans are. They pour all their energy into growth, as a single bean contains the blueprints and energy needed to turn the sun's light and the nutrients in soil into shoots and leaves. If they have a purpose—and all living things need to have a purpose—it is to make more beans. Beans are, after all, their past and their future. Their children. The next generation. That is pretty much the entirety of what they do, and insofar as such a simple, subsentient living creature has an identity, that's who they are.

So as summer wanes, I mark certain beans with tape—particularly my largest and most vigorous plants. Those beans I allow to grow and grow and then to yellow and wither. When they dry out, I pop them from the plants. Each of those dried beans yields three or four perfect seeds, utterly indistinguishable from those that tumbled out of a packet I bought years and years ago.

I let them dry further, and then they all go into a little jar, which I'll seal up with some moisture-absorbing material. Seeds enough for fifty or more plants inhabit that jar over the winter. On the one hand, that's the practical thing to do and will save me a couple of bucks come next summer. Why pay for seeds when your plants will give them to you for free? My Scots blood burbles happily in my veins at such a delightful opportunity for thrift.

But thrift is only part of the motivation for saving seeds. I feel a peculiar but inescapable gratitude to these simple living things. I've tended them, nurtured them, and cared for them. I know that their entire purpose is to have more beans, their offspring, grow next year. In exchange, all summer long, they've fed me and my loved ones. I know that more beans are the simple purpose of those simple things. Having gotten to know them, it seems the least I can do is help make that happen. It's that level of awareness—a deep and fundamental gratitude for what God does for us and for the harvest—that we will need to develop as we move toward a life on a world where the harvest will be even more hard-won.

Nurturing that sense of attentiveness to creation is a large part of the reason why the church I'm now serving created that community garden on our property. Twenty-four raised beds filled with good, local soil sit in a lovely, landscaped area surrounded by indigenous plants. In those beds are greens and strawberries, peppers and tomatillos, zucchini and tomatoes, all grown by a mix of church folks and residents of our little town. Kids from the local schools come and plant. There are a few higher raised beds, which are better for aging backs and knees, for seniors from the local senior center. That garden connects us to community, and to one another, and to the reality of what it means to produce food.

When all of life is at a remove, distant and part of processes to which we feel no sense of connection, we forget who we are. We lose ourselves and our souls in the rush of mechanized, distant processes, and we forget the potent graces to be found in more intimate spaces.

That powerful need to reclaim the intimate is why I have devoted so much of my life to leading small faith communities. On the one hand, it seems like a fool's errand. What possible relevance can small groups of church folk have to the great challenges that face our world? What can a deep dedication to local relationships have in a time of global climate crisis?

Our yearning, when faced with a problem of this scale, is to go big. Think corporate! Think growth, growth, growth! Small churches don't offer that. They are an old and deeply human way of being together. But they're not reflective of the dynamism or growth or profit ethic of our consumer culture. They are out of step with both our sprawling globalism and our deepening ability to virtually surround ourselves with exactly the folks we want to be with.

If you don't like a faith community in a consumer culture, your remedy is simple. You just leave it. As a faith consumer, you find another brand that better suits your interests. If the pastor preaches something that isn't exactly what you think is true, or if someone does something that steps on your toes, you just go somewhere else. Go to another church that better suits you. Or stop going to church at all. It's your choice. We're all free to leave, thank the Maker. Find the place that is exactly right for you, our society says, and so we do.

That can be a good thing on so many levels. Being forced to remain in an oppressive community is a nightmare. Being forced to stay in a place where you cannot be authentically yourself is a terrible thing. Staying in a church where you aren't growing is a mistake.

And small takes work. The work of seeking consensus, the mutual forbearance, and the patience necessary to sustain the life of little churches? These can be hard, particularly if you feel passionate about X or have found your life's purpose in Y. It is much, much easier to seek out the ideal: the community where X is everyone's passion and everyone around you believes Y.

You can't do this in healthy, small churches. You just can't. There, kindness, patience, and forbearance must rule. A willingness to show grace in authentic difference has to abide, or the whole thing comes apart. Or it devolves into unhealthy things, closed off by groupthink, controlling, bitter and shallow and broken.

I can see the relevance of the small church to a healthy family life and relationships. It bears a strong resemblance to those things. A willingness to live graciously with difference and not seek your own interest above your partner's life is a vital part of any marriage or relationship. The same is true in the church: power and self-seeking can tear it apart.

But in the "grand scheme of things"? I've struggled sometimes. In my weaker moments, tiny churches seem quaint and irrelevant cast against the grand, bright scale of our world, where power and profit and growth and ideology rule. It feels weaker still, on first blush, against the epochal climactic changes that are sweeping over our world.

Then, out of some deep recess of my subconscious, I remember that famous image of a little blue dust mote. It's a picture taken by the Voyager spacecraft as it journeyed forever away from our world. Just a tiny, bright, azure dot, in a field of space dust, a dot that just happened to be our planet. All we know and everything we are exists in a tiny, limited space.

We are creatures of a humble planet, just one tiny particle in space. And we can't leave, not yet, not in any meaningful numbers and not for any significant period of time. When we imagine that the virtual worlds we create for ourselves are reflective of our reality—those places where we surround ourselves only with People Like Us—we're deluding ourselves. When we surround ourselves with like-thinkers, the hum of that echo chamber that sounds so comforting in our ears is a falsehood.

This world is itself a small community, a little tiny island in a vast and inhospitable ocean. There is nowhere else for us to go. We can't just pack up and storm off because of our passion for X or our belief that Y is the one true way.

And suddenly, the learnings about what it means to live graciously in smallness are relevant again. Humility is not the smallness that is oblivious to its place in God's creation. Humility turns our eyes to the dirt below our feet and the neighbors around us. Humility means that we live not siloed away in our virtual echo chambers or cosseted in the buffed falseness of brand but among neighbors. Humility connects us.

We have to be aware that we are connected because we *are* connected. We're here together on this tiny, tiny world, a world that we have in our foolishness managed to turn harsh and hard against us.

CHAPTER 15

BECOME A VEGETARIAN

Everything is connected, or so the saying goes. As much as that's become a truism, it remains something worth reminding ourselves of every now and again.

We hum along to that old Beatles song (at least those of us who remember who the Beatles are), knowing that in the end, the love we take is equal to the love we make. We hum along to the *Lion King*, which we watched so many times on DVD when our parents needed an hour of quiet time, about the circle of life, how it *ruuuules us aaaaalll, through despaaaair and hooooope, through faith and luuuuf.* It's a thing we've heard so many times, we don't even think about it or what it means.

We forget that truth—we really do—in the way that a people do when everything is neatly divvied up and packaged for us. Oh, we *think* we're connected. We think we're as connected as human beings have ever been. Here we are, in an age when we can see whatever we want, and have whatever we want, as long as our credit is good. It arrives two days after we click a button on our trackpad, neatly Bubble-Wrapped in a brown box on our doorstep. Anything at all, anything in the world, right there whenever we want it. Want a GPS mount for your new motorcycle? Boom. A vintage analog synth? Any book ever printed, any movie ever made, seeds to grow anything? We can have it right now.

And yet the reality underlying that sense of instant industrial interconnectedness is that we are personally oblivious to the material things that flow through our consumer lives. We do not build them. We do not see them built. We do not know what goes into their production. They just sort of appear, as if they have been conjured up by some peculiar magic. This can be neat, I suppose, but in some circumstances, it is rather less so. Like, say, with the stuff we eat and, in particular, the creatures we consume.

I was reflecting on this reality on a recent afternoon on Father's Day, as I sat by the grill and prepped the food for my gathered family. I love grilling: love the primal character of it, love the fire and the smoke and the scent of it. I remember, hard as it is being a vegetarian, that the reason Cain was so mad at his brother Abel was that the I AM THAT I AM liked the smell of sizzling meat more than the smell of veggies.

There my little soy protein circlets and zucchini cutlets sit, grilling away, as the rich, flavorful incense of burning beef fat and barbecue sauce mingle and rise to the heavens, and I have to give Yahweh that one.

Still, it's a little weird for a vegetarian to be out there at the grill, cooking up chicken and burgers, because there's something I know. Those boneless thighs and breasts, those sizzling patties, they all came from somewhere. They did not magically appear, neatly plastic-wrapped and prepared, having been vat-grown in some huge facility.

And as I cook, and that scent rises up to heaven, I think about the formerly living beings that I'm cooking, about the specifics of what they are and were. I consider their existence, their lives, as known to God as my own.

Having read my Bible for the last forty years, I wonder at how my connection to the creatures we consume plays off against what I know from my faith about my bond with all beings. I accept, for example, the ancient teaching from the book of Ecclesiastes. Like the book of Proverbs, which lays out the essence of wisdom, the point of all ancient Hebrew Wisdom literature is to show us how to live in ways of quiet, ungrasping righteousness. Ecclesiastes is traditionally ascribed to Solomon, Mr. Big Wise Wisdom himself, the most discerning of the kings of ancient Israel, but the book itself seems to draw provenance from a later period. The book is coy about authorship, telling us only that the author's name is Qohelet.

In the Hebrew, that means either "The Teacher" or, more exactly, "The One Who Assembles." That works for teaching on so many levels, frankly.

So words come to us from the Teacher, the author of that taut little book of deeper wisdom.

Much of what we get from Qohelet is complex and challenging. On the one hand, Qohelet is the great cynic of Scripture. He knows that where Proverbs claims that the righteous prosper, being good often yields nothing but suffering, and the evil can do quite well for themselves. "Meaningless, meaningless, all is meaningless," he sighs, in an accent that could be French, while leafing through his dog-eared copy of *The Stranger* and furtively smoking an unfiltered Gauloises. But while it's intended to be a stone-faced reflection on the mortality of humanity—of how we are but creatures of earth just as the animals around us are creatures of earth—I've actually found it to be one of my go-to comfort passages in a very particular instance.

We all have loved animals—at least, the ones we've gotten to know. And when that tiny, shy, little puppy grows up into a reserved dog, we don't love them any less. And when that old cat dies, we feel the loss. They're family, and we don't just say that—we feel it. We mean it. As we consider their souls, from the heart of Qohelet's wisdom comes this passage: "I also said to myself, 'As for humans, God tests them so that they may see that they are like the animals. Surely the fate of human beings is like that of the animals; the same fate awaits them both: As one dies, so dies the other. All have the same breath; humans have no advantage over animals. Everything is meaningless. All go to the same place; all come from dust, and to dust all return. Who knows if the human spirit rises upward and if the spirit of the animal goes down into the earth?'" (Ecclesiastes 3:18–21 NIV).

So when I am asked by mourning children or sorrowful adults what happens to our pets after they die, I can point to this passage and say from the deepest wisdom that their breath is like our breath. They return to God, just as we do. And we can take comfort in that.

But then I also know a truth from elsewhere in the great story of Scripture, one that hums oddly when I try to harmonize it with that truth. That other truth comes from the very heart of the teachings of my own Teacher, from Jesus himself as his story is told in Luke's Gospel.

It's from a section of Scripture known as the Sermon on the Plain, which is the shorter, tauter version of the Sermon on the Mount that appears in Luke. Where Matthew gives us three full chapters, Luke encapsulates it in twenty-nine verses, starting in verse 20 of chapter 6 and running through verse 49. All of the core teachings are there, with the radical moral imperative to show grace, mercy, love, and justice to all as the sure foundation under all of it.

It is an immense and fiercely challenging teaching, calling those who'd follow Jesus to commit themselves to proactively seeking healing. The Way that Jesus lays out is the path of transforming love. The Christ-follower is asked not just to passively receive hatred but to push back hard against hatred with grace and mercy—to double down.

Right here in these words is the moral core of all Christian faith, the ethic that guides how we are to act in this world. The words in Luke 6:38 sing out an image—a feeling, really—of what God's justice looks like: "A good measure, pressed down, shaken together, running over, will be put into your lap; for the measure you give will be the measure you get back."

It's that truth again, only less passive. Jesus presses it down to intensify the flavor, mixes it up, and then pours it out over us. We are connected to all things, and our every action will be returned to us.

Which is why, when I look at the reality of how we eat and the reality of how most of the meat we consume gets to our grills and our tables, I'm a little reluctant to be a part of that reality. Because most of our meat—the remarkably affordable packaged bulk meat that fills the fridges at Harris Teeter, the meat in our fast-food burgers, the meat in our *banh mi* sandwiches—comes to us through the miracle of modern mass production.

When I was a boy, I remember taking field trips to factories. On one tour in sixth grade, we watched musical instruments—including trombones, trumpets, and tubas—being manufactured. There were big machines and workers and hammered brass, and it was really cool. Years ago, when my wife and I celebrated our fifteenth wedding anniversary with a week of Vermont bed-and-breakfasts, we stopped in at the Ben and Jerry's factory and took the tour. We saw every step of the process and got to taste some weird new flavors. Snozzberry Crunch, or something similar to that. That, let me tell you, was pretty awesome. Those factories stirred in me an appreciation for production. I felt proud to be part of such a thing, in my small way.

But a tour of a modern factory farm or industrial-scale slaughterhouse would not feel quite the same. I don't think Mr. Ramsey's sixth-grade class would ever quite recover.

We are part of the systems that sustain us. Their reality, shaken together and running over, is the measure we will be given. What these passages, in conjunction, ask us to consider is the depth of our connection to the creation of which we are a part and over which we are meant to exercise care. As we move further and further into systems that isolate us from the reality of how things are made, it's important that we keep the knowledge of our connection front and center.

I'm certain, in fact, that I'm right about this—so certain that there's a danger to my soul. The more sure we are of our righteousness, the more risk there is that we'll become self-righteous and smug about our own correctness.

Having been a vegetarian for most of my life, I'm at great risk of smugness, and that threat has only grown deeper as the climate crisis has become more evident. The threat was only deepened by the news of a recent study by Oxford University. Not just any study but an Oxford study, meaning it has its provenance in a university that will soon be celebrating its one thousandth anniversary. This, in and of itself, seems grounds for smugness. ("Oh, your town just turned one hundred and fifty? How lovely! I remember my alma mater's one hundred and fiftieth. It was eight hundred and fifty years ago.")

I know my diet is better for the environment because that is precisely what the study at Oxford says. A vegetarian diet uses less than half of the land and produces half of the carbon emissions than a meat-based diet does. If we were all vegetarian, we could save the planet, or so the headlines ran. Seriously, how much more smugness potential could exist in a dietary form?

The study itself added significantly to my dietary smugness potential, for as a vegetarian, I already find it so very easy to feel superior. It's been so long since I stopped eating meat that I've kind of forgotten when I stopped. It was sometime after I got married. That, I remember. So when anyone asks, I'll say, well, I think it's been eighteen years. Sometimes I say fifteen years. Other times I say twenty.

I know a diet of vegetables is healthier—of course it is. There is nothing healthier than being a vegetarian. Now, one might say, it's too hard to be vegetarian. What can you possibly eat? But it's easy to enjoy this healthful diet, I say. Take, for example, pizza. Pizza is vegetarian. Pepperoni, not so much. But pizza? It's fine. Beer is also vegetarian. As a pro tip, I will note that New York Super Fudge Chunk Chip ice cream is also vegetarian. This remains true whether you eat but a spoonful or manage to down the whole pint. Whichever way, it's vegetarian. See how easy it is to be virtuous?

And these days, it's even easier to be vegetarian than it used to be. There are countless more things to eat, up to and including nifty new veggie products that replicate the flavor and texture of meat.

I know it's kinder. We get our meat from hideous factory farms, and even if we don't, poor Bambi is still crying alone in the forest. "What's wrong?" asks Thumper. "Someone wasn't a vegetarian," says Bambi, at which point even more smugness arises.

For all of this, it's not that I'm lacking conviction that vegetarianism is better or that a plant-based diet is less vital to our future. In the face of all of this, I find the apostle Paul's discussion of a form of early

Christian vegetarianism so appropriate. Here though, what's both strange and worth noting is that the smug folks weren't the vegetarians. They were the carnivores. Paul, an omnivore himself, describes vegetable eaters as "weak," which is totally unfair because, gosh darn it, *I'm* supposed to be the self-righteous one.

The issue for early Christians who chose not to eat meat was not that they were worried about climate change. Instead, the issue lay in consuming meat that had been sacrificed to idols. This generally isn't a concern when we stop by McDonald's or Red Robin or White Castle, but back in the first century, it was a thing. Meat in the highly dynamic, pluralistic culture of the Greco-Roman world was often . . . there's no delicate way to put this . . . "used" meat. That meant that before it went for sale in the marketplace, the animal involved had been sacrificed at the altar of one of the almost countless gods of the ancient world. While a small amount of the sacrifice would have been burned, most of the rest of an animal would have been either (1) consumed by the priests or priestesses of whatever god it was sacrificed to or (2) sold to the market as income for that particular temple.

For some early Christians, this was a major issue. They'd just converted to the movement that worshipped Yeshua ben Yahweh, and they knew that they were supposed to worship only one God. Having rejected all other gods, they were terrified that they might somehow be violating their relationship with Christ and their Creator if they had some BBQ ribs that had first been sacrificed to Ba'al.

Rome, standing as it did at the very heart of the empire, was filled with temples and altars. It was chock-full of ancient religions and mystery cults. Some of the fledgling Christians in the town feared that they might accidentally lose their Jesus connection if they ate pagan meat.

For those who were wiser and more well read, the whole idea seemed absurd. And it was to those others that Paul—who shared their perspective—directed this section of his letter. Paul, though a stranger to the Roman church, knew it by reputation. The church in Rome comprised, or so we can assume from his writings to them, the erudite: philosophers, the wise, the aware, and those steeped in the knowledge of how things really were.

Paul is not rejecting them. In fact, Paul is showing that he believes exactly the same things. For those whose grasp of the faith was strong and who understand they are free to eat and act and live in ways that stand beyond the grasp of others, Paul is clear: don't be judgmental. Remember, he softly chides, that there are things all of us disagree about. Should we celebrate some of the sacred days of the Jewish tradition or not? Some think yes, and some think no; what matters most is that we do what we do in a way that honors and respects those who do not share our position.

What Paul does not share is their willingness to condemn or mock those fledgling Christians who lack the depth of understanding that he shares. He acknowledges that they are "weak," sure. But what he will *not* do is act or live in ways that subvert the faith the weak do have. If you love others, you don't live that way. Possessing knowledge is not enough. You have to use that knowledge in ways that do no harm.

The task of every Christian—even in disagreement, even when you know you're right, even when your absolute correctness is utterly and empirically provable to any halfway sentient being—your task is to love, to build up, and to do no harm. Your task, as a follower of Jesus Christ, is to bring no harm to another.

Ultimately, that was the apostle Paul's measure. "Therefore, if food is a cause of their falling," he says, "I will never eat meat, so that I may not cause one of them to fall."

I am convicted, from both science and the defining compassion of my faith, that the consumption of meat brings harm to the world. I believe that in ways that now go far beyond my desire not to participate in the suffering of animals. On our new planet, it doesn't matter that meat smells delicious and tastes even better. What matters for my soul is that my appetites do not contribute to the hunger and thirst of others. I act on that awareness by doing what Paul did. I don't eat meat, right now, because to do so is to place greater demands on our planet than can be sustained.

As our world grows harsher, and arable land becomes less abundant, and it becomes harder to feed humanity, that diet bears more moral weight. Sure, I might like the succulent flavor and primally satisfying texture of a nice bit of steak, perfectly marinated, sizzling on a grill. I mean, seriously. I'm salivating writing this.

But if we know that this choice is contributing to the harm of those more vulnerable than ourselves, then acting on that desire is problematic. If we know without question that when millions of us make that choice, tens of millions will go hungry, we just can't. Not because it's inherently monstrous—we are omnivores by design—but because we are moral beings blessed with the ability to discern the impacts of our desires and appetites.

As a higher primate, I have many desires that are woven into my design. I desire to reproduce, which manifests in a generalized attraction to females of the species. That's a desire that has only marginally lessened since the fires of pubescence first lit that fuse. But I contain that energy within the bounds of my attraction to my wife and within the deep and lifelong covenant commitment I have made to her and to my family.

I desire to have power and social standing, a hunger that arises from countless millennia of life within primate troops and clans. Status ties in with reproduction and health and well-being, as those with status have always thrived. To maintain our place, we snarl and we bite and we attack. I feel that old passion just as strongly as any, particularly when spending time on Twitter. But I am more than that, and the commitments that give me my integrity go deeper. So I don't attack my enemies. I don't strive for dominance, as much as those yearnings stir in me.

The longing to eat meat is like those other very natural things. I may feel the urge to copulate outside of covenant or the urge to kill my enemies, but I don't. Knowing now, as I do, the harm of eating meat and the cost of it to our world and our future? I simply can't do it. As a Christian, I am not to be ruled by the flesh.

As our faith stands in encounter with the ethical demands of a new and harsh world, the choice to move to a plant-based diet will need to become a significant part of a Christian identity.

CHAPTER 16

RENDER UNTO THE REPUBLIC

On a recent Tuesday, I had a job to do. That job was simple: vote. That particular Tuesday was not a presidential election and not an election for the United States Senate or the House. The stakes were more local and more intimate. It was a state and local race for county supervisors, school-board members, sheriffs, and state senators and representatives. This was small-scale, rubber-meets-the-road democracy, where much of the nitty-gritty labor of our community life gets done. In the last weeks before that Tuesday, the medians and roadsides flowered with hundreds upon hundreds of wire and plastic signs proclaiming the names of those who were competing to lead us.

For all the sound and fury and star power of our billion-dollar national election industry, state and local elections are where the choices actually get made about how we're going to run our schools. About how we're going to keep our communities safe. About where our priorities lie.

As I sidled off my motorcycle and walked into the little white Episcopal church on a hill that serves as my neighborhood polling place, it became obvious that our priorities lie elsewhere. There was the usual diligent cadre of retiree volunteers. There were the two party volunteers parked out front, handing out voting guides. There were the voting machines, their touchscreens glowing softly behind their privacy shields.

But there was no line. In fact, there wasn't a single other citizen in there to vote. Not one. I breezed right in. I breezed right out. Wow! That was so convenient!

That was so . . . wrong.

After conferring with my wife, I learned that it wasn't quite as effort-less and empty in the morning, but there was still only a very short line. It was less crowded than a candy store the day after Halloween.

How many people bothered to vote? On Wednesday morning, I went online to the Virginia State Board of Elections website. I looked through all of the statistics. If we believe our own propaganda, America is sup-posed to be one of the great cradles of democracy. Virginia was one of the thirteen colonies to first ratify the constitution of the United States of America. But on that particular Tuesday, only one in three Virginians bothered to vote. That's 34 percent. Thirty-four percent isn't just a failing grade. It's the kind of F you get when you take your Political Science 342 final exam after first downing a fifth of bourbon and several grams of psychedelic mushrooms.

Now there are lots of reasons why we don't participate. Maybe we're too busy. Maybe we just didn't have time to decide. Maybe we don't really like the political position of either of the parties. Maybe aliens took us from our beds on Tuesday morning and, despite our protests, weren't done with their experiments until after the polls had closed.

Are those really reasons? Or are they just excuses? But elections aren't important, you might say. They don't really change anything. It's just those fat cats in Washington lining their own pockets. Again, those aren't really reasons not to be engaged. Excuses, yes. Rationalizations, sure. But reasons? No. Cynicism is the mask worn by those who wish to justify shirking their duty.

Well, you might say, fishing a bit, Jesus never told me I needed to vote. You are on to something: indeed, Jesus never voted at all! Find me the word *vote* in the Bible! Ah-hah!

Jesus actually talked a great deal about how we should respond to gov-ernment. The leadership in Judah had their eye on him as a troublemaker and a rabble-rouser. In fact, they were so eager to get this guy off the streets that they tried to set up a sting operation to trap him. It was going to be tough. He was popular, and they were going to have to be crafty about it. We can read about that effort in Matthew's Gospel (Matthew 22:15–22).

Rather than arrest him themselves, they tried to get him in trouble with their imperial Roman occupiers. Their agents asked him about whether Jews should pay taxes to Rome. It was a cunning trap. On the one hand, if you answered yes, it meant that you were willing to use Roman money, on which were inscribed assertions of the emperor's divinity. It meant that you were assenting to him as a god and thus betraying the God of Israel. It also meant you were supporting the hated occupiers of the Holy Land. So you couldn't answer yes, or you were a traitor to the Jewish people.

On the other hand, if you answered no, it meant that you were a dangerous revolutionary, a threat to the empire. The Roman authorities didn't look kindly on people who refused to pay their taxes. So you couldn't answer no, or you were a threat to Rome.

Jesus was not so easily taken in. He didn't say yes. He didn't say no. He just told everyone to look at the coin and see who was on it. It was the emperor, of course. So give him what belongs to him, and give God what belongs to God. Render unto Caesar what is Caesar's, he said. It was a perfect answer: both yes and no and neither yes nor no. And the trap they had set for him snapped closed on empty air.

But as we hear his answer, we have to ask ourselves, What is it that we owe the "emperor" today? What do we owe to the emperor when we, the people, are the emperor? Not just our taxes. This is a democracy, and what a democracy needs from its citizens in order to thrive is participation. What our democracy needs from us is for us to pay attention, for us to be engaged. When we fail to do that, we fail to give to Caesar what Christ told us is his due. We need to hear that passage in that way in our lives as citizens of our counties, of our states, and of our nation.

Give to the emperor what is the emperor's, Christ said, and give to God what is God's. We, as citizens of this great republic, need to give it our engagement and our participation for it to thrive.

That commitment to the responsibility and power of citizenship was mirrored and affirmed by the apostle Paul. Paul did not shirk from the mantle of citizenship or its power. When he found himself imprisoned for spreading the word about Jesus, Paul stood on his rights to the point of challenging his jailers. "I'm a citizen," he said. "I have rights, and you need to apologize for unjustly imprisoning me." Which his jailers did (Acts 16:35–38).

When he found himself threatened with mob violence, Paul used his citizenship to maintain his right to speak. When Paul was threatened with a public beating, he declared his citizenship loudly and directly, challenging the power of Rome not with defiant indignation but by standing on the most fundamental assumptions of Roman social order (Acts 21:27–40; 22:22–29).

In the thirteenth chapter of his letter to the church at Rome, Paul described the duties of a Christian toward the culture they inhabit. Give what is due, Paul writes, echoing the words and teachings of Jesus. In an empire, again, all we owe is taxes and obedience. Paul was willing to respect the dynamics of that system of governance even with the realization that it might one day take his life. (It did.)

Paul's exhortation to the duties of engaged citizenship brings us again to how we are meant to live as Christians in a constitutional republic. As tempting as it might be to hide away, or to demur from the social contract, or to reject it as corrupt, both Jesus and Paul teach against our abandonment of our responsibility to our nation and to one another.

That truth is not changed because we now face a time of national testing and challenge.

We are in a time of crisis, and the choices we make about our national direction in a time of crisis define our souls. If you believe that government is always and inherently corrupt and incompetent, you will expect corruption and incompetence. Through apathy and inaction, you'll materially tolerate corruption and incompetence. You might in fact help elect corrupt, incompetent leaders through both your cynicism and your disengagement.

Faced with the reality of a planet that is becoming harsher and less forgiving, one that will inflict deepening suffering on humanity for generations, we can choose willfully obtuse, ideologically blindered leadership. We can choose leadership that refuses to accept the reality we have created for ourselves. We can vote in representatives and senators and administrations that tell us the lies that we want to hear. "Everything is fine," they say. "There's nothing to worry about. It's all fake. It's just weather. This climate change thing is just a hoax."

We can vote in leadership that values and listens to short-term profits and padding quarterly margins of major corporations over the interests of our children and our communities, that protects the portfolios of the investor class and hedge-fund managers over the needs and well-being of the average American. And because we're dealing with an event that will play out over multiple human lifetimes, it'd be easy to listen to their siren song.

It's a siren song that destroys international cooperation in response to our rapidly changing world. It's a siren song that encourages us to double down on our consumption of remaining fossil fuels and prevents us from beginning the response that might save us from the worst possible impacts of what already has become a significant and catastrophic reality.

And most significantly, for our souls, it's a siren song that lulls us into personal moral complacency, allowing us to gloss over our individual responsibility for our actions. How we choose is shaped by the leadership we choose. And unlike in an empire, where we can blame the oppression of an unrepresentative hereditary leader whose power rests on fear and coercion for the ills of our society, our republican form of government places that responsibility heavily on our shoulders.

Failure to act on our duty to lead our republic is the wrong choice. It deepens harm both to us and to the generations to come. But again, the

Creator of the Universe is willing to allow us to make the wrong choices and experience the consequences of our wrong choices.

Faced with this reality, Christians living in a republic are morally obligated to engage politically. As individuals who have chosen the path of Christ, we are called to engage with the governance of that republic. We use our vote as a way of shaping the spirit and moral direction of a nation. We know, with a high degree of certainty, that voting in certain leaders will result in foot dragging, denial, and a willful failure to act.

It will result in continuing to put short-term profit and gain over the future of your children and your grandchildren.

It will result in a fundamentally anti-Christian attitude toward the stranger and the refugee. Manipulative charlatans will malign desperate people fleeing a crisis for which we are responsible as dangerous, lazy, and a threat.

As a basic metric, morally and ethically, our duty as persons who are both Christian and citizen is to use the God-given blessings of liberty to move our nation in ways that do not harm creation or our neighbor. You know what that looks like. Acting on it is an obvious moral duty. But that doesn't mean that we, as citizen-leaders of a republic, will act. Because those in power always have a special place in their hearts for the voices that whisper to us the things that we want to hear.

The prophet Jeremiah knew all about people who told the sweet lies that power wants to hear. Following the destruction of the Assyrian Empire in 627 BCE and the death of its last emperor, Ashurbanipal, the people of Judah had hoped that they would finally be free of imperial oppression. Judah and all of the other nations that had been enslaved by Assyria rose up in revolution. Led by the wise and noble king Josiah, the people of Judah reestablished worship of the God of Israel and hoped for independence. But it was not to be. In 609 BCE, Josiah was killed by the Egyptians at the battle of Megiddo, as the Pharaoh's army raced up to aid what was left of Assyria in its struggle against the new power that was rising in the region.

That power was the Babylonian Empire. Judah found itself enslaved again under a more brutal master than before. All its efforts to rise again were brutally crushed until, in the year 587 BCE, the Babylonians finally destroyed Jerusalem completely, tearing down the temple and scattering the people to the four winds.

Jeremiah lived and preached in those last, terrible days before the final Babylonian destruction of Judah. He was not a popular man in Judah because he proclaimed that to resist Babylon out of national pride would result in complete destruction. The visions he received from God were relentlessly negative and challenging. At best, he was seen as a prophet of doom, a weeping prophet, a proclaimer of despair. At worst, his fellow Judeans saw him as a traitor. How dare you undercut us? How dare you tell us that God will not always support us no matter what we do? He was imprisoned. He was thrown into pits, actually and physically. His life was threatened.

Standing against him were other prophets, teachers, and visionaries who presented the leadership of Judah with a completely different vision of the world. Judah was the nation most blessed by God, they said. God would protect Judah no matter what happened. Those prophets described a future of endless prosperity, a future in which Judah would rise from one height to the next. Everything the leadership did was wonderful and amazing, and God's blessings would pour down upon them. They taught that Jerusalem was Zion and that Zion, as the seat of Yahweh, could never be harmed.

Zion was inviolable, was blessed, and could not be harmed.

What these prophets of prosperity and power declared was a sparkling vision of glory. It was also completely and utterly wrong. That, unfortunately, is the challenge that faces us whenever we want to believe propaganda and self-serving nationalist rhetoric rather than the hard, fierce reality of God's work around us. Just as the dream of Judah's power fell apart under the weight of its own lies, any nation that would rather hide in the shadows of falsehood will soon find itself shattered by the reality of a creation that will not yield to our desires.

"What has straw in common with wheat?" rumbles the voice of the Lord in Jeremiah. "Is not my word like fire . . . and like a hammer that breaks a rock in pieces?" (Jeremiah 23:28–29).

A nation run by a proud fool incapable of hearing anything that challenges him will fail. A republic run by millions of proud fools will fail just as surely. It doesn't matter how stubbornly we cling to our false assumptions about climate change or how loudly we shout that it's a "hoax." If we don't use our freedom to act in accordance with what is true and adapt personally and as nations?

Well, God does what God does.

It's not like we haven't been warned. And as we vote or choose to shirk our duty, we can expect those choices to be weighed against our souls, both personally and collectively. God is just, after all.

But against the terrible truth of God's justice, there's something more.

CHAPTER 17

EMBODY GRACE

Few verses in the New Testament are more familiar than John 3:16. It's at the very heart of the Evangelical movement. It's meat-and-potatoes stuff—or perhaps, given the diet we'll need on this new world, tofu and potatoes.

Here it is: "For God so loved the world that he gave his only Son, so that everyone who believes in him may not perish but may have eternal life" (John 3:16).

It's a quick, catchy verse that sums up, for the most part, the whole purpose of Jesus. It is also a shorthand verse that makes its way onto an array of Evangelical products. You can buy John 3:16 caps. You can buy John 3:16 shirts. There are bumper stickers in an endless array of different colors. There's John 3:16 candy corn. There are these little John 3:16 key-ring bells they market to bikers, which tinkle quietly as you motor along. Also for bikers are some John 3:16 pant clips to connect your jeans to your riding boots. Because God so loved the world that he didn't want your pants to hike up your legs in an unattractive way as you motor down the road on your Harley.

Before there was Tim Tebow, with John 3:16 inked in his eye black, a guy known as Rainbow Man helped popularize the verse. Waaay back in the 1970s and 1980s, Rainbow Man showed up at pretty much every sporting event he could wearing a huge, multicolored afro wig and waving a large, handmade sign. That sign, often written on a simple sheet, said "John 3:16." Christians knew what that meant, and the idea was, outside of Rainbow Man promoting himself, getting people curious as to what John 3:16 actually was.

The man's name was Rollen Stewart, and every year he'd rack up more than sixty thousand miles of driving as he trucked himself across the country. Every Sunday, during at least one major televised sporting event, there he'd be, waving his homemade sign. He became somewhat-sort-of-kind-of famous, the nation's most well-known sports fan. He also became a small celebrity in the Evangelical Christian community.

And then he disappeared. Well, he didn't so much disappear as he ended up convinced that he had to tell the world that Jesus was coming back to destroy it soon. It was 1992, and the end times were about to be fulfilled, and everyone needed to repent or be destroyed in the very near future. As no one seemed willing to listen to this less cheery message, Mr. Stewart took three people hostage at gunpoint, demanding that police give him access to the media so that he could tell the world that destruction was imminent.

After eight and a half hours of armed standoff, Mr. Stewart was apprehended by the police. He's now doing three concurrent life terms at Mule Creek State Prison near Sacramento. (He's eligible for parole in 2022.) In an interview with the *Los Angeles Times* given a decade ago, he shared that he's unrepentant, convinced that he only botched the timing of his message. He remains absolutely positive that he did the right thing in trying to warn people that the world will soon be destroyed. If he had it to do over again, he said, he would. What could have been a simple message of grace became somehow obscured by a growing obsession with apocalyptic destruction.

It's a strange tension for many Christians that can be reinforced if you bother going beyond branding and slogans and actually look at the passage that surrounds that iconic verse. It's a challenging one.

The verse itself comes in the middle of a conversation that Jesus is having with a curious Pharisee named Nicodemus. Nicodemus truly struggles to understand what Jesus is talking about, as he's prone to taking things—like being "born again"—completely literally. As Jesus tries to explain himself to Nicodemus, the conversation turns to the purpose of the coming of the Son of Man, one of the terms that Jesus uses to describe his relationships with God. What is his purpose? His purpose is defined by God's love for the world and by God's desire to give those who know the truth of Christ's grace eternal life.

That's the 3:16 part of John. But Jesus may not be finished talking. Or maybe he is. We aren't sure, actually, where the words of Jesus end in this passage. That is in large part because the Greek language in which this Gospel was written doesn't happen to have anything like quotation marks. You just have to kinda figure out from the context of a sentence when someone is no longer speaking. Many reputable scholars think that Jesus is done in verse 15 and that verses 16–21 come from the voice of the Gospel writer.

Either way, the passage goes on, as conveyed by John's elegantly simple Greek, to describe the purpose of the Son of Man further. Here we start to get some mention of condemnation and judgment. Verses 17–19 talk a great deal about who is and who is not condemned. Coming off of the grace of verse 16, the rest of this passage seems to stand as a stern warning: believe, or be condemned. The transition can seem as intense as the one that was expressed in the life of Rainbow Man. Grace on the one hand; threat on the other.

Here, though, it helps to know a smattering of Greek and to be aware that sometimes what is expressed in the gospel we read may not be exactly what was written by its author. English and Greek are related, as English relies on Greek for many of its words, but they are not the same.

The word *condemn*, for instance, has a strong negative connotation. You condemn a building when it's time to knock it down. You condemn a rainbow-afro-wearing, John 3:16–quoting hostage taker to prison for pulling a pistol on bystanders and the police. Condemnation means a judgment in the negative. The problem is that the word that the most ancient texts of John's Gospel use to describe what the Son of Man is doing in verses 17–19 doesn't have that negatively charged meaning. The word is *krino*, and it just means "to judge" or "to decide." It can just as easily imply a positive outcome as a negative outcome.

Where it appears in verse 19, for instance, as the related word *krisis*, any sense of being a negative value judgment is completely missing. It is simply *judgment* or, if you're using the NIV, *verdict*. Verse 19 would read pretty strangely if we translated it as "And this is the condemnation, that the light has come into the world." That would be as far from John's point as it's possible to get; that point is the depth of God's love as expressed in the person of Jesus of Nazareth. What we get when we go back to the original language is a much more consistent passage—one that speaks primarily of God's love and God's willingness to embrace humankind and one in which condemnation and damnation are not painted as a central part of what Jesus is all about.

There is a core gospel challenge for Christians, I think, in this time of crisis. While grace is at the very heart of the teachings of Jesus, it's easy for the message of grace to be lost in our anxious, grasping desire to be sure of our own righteousness. It's easy for the message of God's kingdom to be lost as we dither endlessly over who gets to go to their ten-thousand-square-foot mansion in heaven and who gets Left Behind. It's easy for us to take this message of reconciling love as just another thing to argue about or to use as a club against those who don't believe as we do.

As the story of Rollen Stewart reminds us, balancing judgment and grace is a challenging thing when we're sure the end of the world is at stake. It's easy to become so focused on the all-caps NEED TO PREPARE FOR THE END TIMES that we set aside a gracious attitude toward others and ourselves.

We look to that friend who bought a pickup truck—a huge, pointless, planet-mangling beast of a thing that doesn't reflect their needs as a CPA—and our heart thrills to the sweet music of self-righteous disdain. We arch our eyebrow at the church member whose house feels as large inside as a medieval cathedral. We see the people at the table next to ours ordering a great, big, bleeding New York strip steak and we sniff haughtily.

Planet killers, we mutter under our breath. *Fools*, we snark. We are better than them. We are more awake, smarter, more caring. And that feels so very good. It makes us feel powerful. It makes us feel like we are the arbiters of all things.

It is also utterly devoid of grace.

From our self-appointed position as judges of all things, our eye finds every last speck in our neighbor's eye. Perhaps they flew in a plane once. Maybe they took a trip to visit their family and they drove a *car*! Perhaps they light their home with electricity that doesn't come from renewable sources. Perhaps they are driving a mere hybrid instead of a plug-in hybrid or a full electric. So we shame them, rolling our eyes at just how imperfect and impure they are. "How can you claim to care for the future of our world," we intone, "if you aren't doing every single last thing exactly the way I think you should be?"

In this strange age of social media absolutism, no one is ever pure enough, perfect enough, or trying hard enough. No one can ever be forgiven or be given the space to learn and grow without heaping helpings of shame. It is powerfully, profoundly graceless.

Similarly, our anxiety around the climate crisis can drive us to have a complete lack of grace toward ourselves. Fretful about not doing everything exactly right, we feel constantly inadequate. We look to this problem that exceeds our individual capacity to influence it. Then, from our uniquely human sense of self-importance, we allow ourselves to imagine that our every single decision is of such vast and momentous import that every morsel that enters our mouth and every single thing we say or do is of infinite and crushing weight.

We fret over our choices and worry that we have not done enough or that we have inadvertently made a decision that will somehow lead to disaster. It's a defining feature of anxiety, which paradoxically makes us feel like *nothing* we do can possibly make a difference and that *everything* we do is so freighted with significance that we fear to act at all.

As we face the slow rise of this great global change, a willingness to show grace, mercy, and forbearance to one another will be a necessary part of the Christian witness. The strange, dark energies that rise from our anxiety and our grasping will do what they have always done in human history. They will lead us to condemn and demonize, criticize and shame, rather than be a beacon of a more gracious and merciful way.

It is far, far easier to do the former than the latter. And in some ways, particularly as you're organizing a political movement, it can be remarkably effective. We get all fired up when we have someone to hate. Not an abstraction, not an idea, but a specific person whom we can revile.

In Saul Alinsky's classic 1971 guidebook for creating social change, *Rules for Radicals*, having an opponent to hate was one of the key principles of organizing a movement. One of the more significant things that Alinsky pitched for those who want to organize for a particular cause is how to approach one's opponents. Alinsky, being a deeply realistic critter, argued that respectful disagreement is absolutely useless when you're trying to motivate a group of folks. People don't get fired up to march and shout slogans if you present them with an honest and balanced appraisal of the opposing position. If you have sympathy for the opposing party—if you see some of the merits of what they're saying and are willing to present their position with all of its nuances and possibilities—then you're a crappy organizer.

Not because what you're saying would be materially incorrect. Alinsky acknowledges that human systems are complex and interwoven things and that even opposing positions likely have positive aspects. In fact, he banks on it, as ultimately his goal is to have his mobilized communities negotiate with his opponents for whatever gains can be made.

But when you're rousin' the rabble, you don't say those things, even if you know they're true. The rhetoric of Alinsky's community organizing is apocalyptic, meaning it is radically binary. Once you've identified your enemy, you define them as 100 percent evil and your own position as the ne plus ultra of virtue and all things good and right and true. When things are pitched out in those binary terms, it becomes much easier to get people motivated. It will be tempting, oh so tempting, for those who are concerned about this crisis to use precisely this method.

Three things strike me about this approach.

First, it requires organizers to be liars. You hold a truth in yourself about your opponent, and you knowingly misrepresent their nature to your own people to stir up passions.

Second, this approach works great. It's wonderfully successful in the political arena. And while Alinsky was writing for the political Left, his methods are far from limited to the political Left. Alinsky's methodology has won some significant admirers on the American Right. When I read smart conservatives—meaning ones who are talking openly about Alinsky with one another for purposes other than faux anticommunist polemic—they like what he has to say. They glom onto his methods. They see how useful he can be. They are now, in fact, using his methods in their training. So far it seems to be working.

Third, this way of approaching one's enemies just ain't Christian. Yeah, I know, Jesus cleansed the temple and took on the powers that be and yadda yadda yadda. But what made Yeshua ben Yahweh such a powerfully different presence was not that he taught us to love and honor our friends and demonize our enemies. That's always been the way of the world. It was that he pressed that love ethic out to include opponents. Yeah, they might be messed up. Yeah, they might be the cause of much hurt and oppression and brokenness. They may be doing things that cause real harm and that are contributing to the catastrophe that we will be enduring for the next dozen generations.

But real, transforming revolution only occurs when you can look at those who oppose you and realize that you've got to love them. Real change only comes when you find in yourself the same Spirit that allowed Jesus to look down from his cross and speak words of mercy and forgiveness.

It doesn't feel as good, sure. It doesn't fill you with righteous, glazed-eye partisan fire. But that's not why Jesus lived and taught and died and rose.

That path, tempting as it is, has never and can never bring about the sense of centered purpose that creates endurance. It causes us to tear ourselves apart, to tear and snark and rip at one another.

And a house divided, as Jesus once put it, cannot stand. We'll need to set that aside if we want to move forward together. Grace calls us to look at the Other differently. It requires us to be authentically ourselves while at the same time finding and honoring the best spirit of those who are not us. Grace opens us to seeing through the haze of our anger and our fear. It helps us find ways to speak so that we are heard and is the solid ground of both reconciliation and repentance.

Is it easy? It is not because grace is not the same thing as acquiescence, apathy, and inaction. If we accept harm done to our neighbors, then we are not being gracious. If we have a heart of violence toward those who oppose desperately necessary change, then we are not being gracious. Grace, after all, is not about me. It's about us.

What does this look like? As the pastor of a small congregation in which there are a range of political perspectives, I'm very aware of the trap of complacent inaction. As an introvert who tends to be conflict averse, I'm doubly aware of it. So when my community moves forward on things, we do so in ways that reflect our best understanding of the best graces of our membership.

I have members I love dearly who are not convinced that climate change is caused by human beings. But they do value efficiency and practicality. They do see the virtue of lower energy bills and more energy independence. I both see and share those values. I can honestly affirm them, and from that place of affirmation, we move forward together.

Have I "won"? No. But the point of grace is not winning arguments or dominance. It's restoration, wholeness, and making things new.

CHAPTER 18

USE EVERY GIFT

Our best possible future in this new and angry Eden requires not just a single approach to survival. It will demand that we find the best of all of our paths. We will be morally challenged as persons and as local communities, but we will also face challenges that will require all of the gifts of large-scale human endeavor. We would prefer, of course, that the "answer" to the climate crisis comes from our own ideological framework or out of the expectations of our own culture. But while the spirit of this age is one of atomization and division, we're going to need to bring every last gift, grace, and skill to bear on the climate crisis if we're to adapt and survive. Just doing one thing in isolation simply won't work.

Like they have for so many in our culture, the dynamics of my life have trended toward isolation. As a pastor, I serve a wonderful community of souls, but my work with this sweet little Jesus tribe is part time, and I don't live nearby. That means I may see that group of people Wednesdays and Sundays, but otherwise I'm working remotely—by myself.

I've also spent the last few years trying to make a go of it as a writer, with modest success. Writing, though, is a semimonastic avocation and also involves a tremendous amount of time spent by oneself. There are those brief moments when you're asked to speak at a conference, or gather with a warm klatch of readers at a book group, or talk to a smattering of souls at a signing event, but those are few and far between.

With that, and with my offspring now adults and my time as a shuttle dad at an end, I find myself in a position familiar to a surprisingly large number of American adults.

I'm alone most of the time.

It's a state of being that's familiar to all too many of us, one that was exacerbated during the COVID crisis. Quarantining and social distancing take their toll on more than our incomes. They can drain our souls. We find ourselves starving for human contact, particularly those of us who draw energy from being in the company of others. For inwardly oriented souls like myself, it's a little easier.

I've always been an introvert. I'm comfortable with—and need—more time by myself than most souls. There are boundaries, though, to what is and is not healthy solitude.

Many days, particularly if my hardworking wife is off on a business trip, it's pretty much just me. If I go for a walk, it's me and an empty suburban neighborhood, one that seems desolate in the middle of what, for most folks, is their working day.

Like many who look around and realize things have become too quiet, I've found ways to get out, creating intentional inefficiencies in my day. I'll do smaller grocery shopping trips, fully aware that I'd rather walk or bicycle to get a small load of groceries simply because it means I'm not rattling around at home. I'll hit the library for books. I don't use the banking app on my phone and instead get out and actually go to the bank. I volunteer, delivering food for Meals on Wheels, which gives me an opportunity to be socially present for the chronically ill and the elderly. I feel their isolation, and taking a few moments to talk and be present is as life-giving for me as it is for them.

And on evenings when I am in an empty home alone, I walk to go get dinner. My target destination: Burger King. Going to Burger King is an old habit, one founded when my boys were small. That restaurant is just over half a mile away from our home, and it was an easy walk for little legs. After swim practices at our neighborhood pool, I'd frequently herd my little guys off on what then seemed to them like a long walk. We'd wander on over to get a cheap lunch, them noodling about and chattering, me strolling along and enjoying some dad-and-sons time.

When we'd get to Burger King, they'd get kids' meals. Being vegetarian, I'd get the veggie burger—which, for years, was pretty much the same mass-produced veggie product that we also had in the fridge at home.

As the years passed, my sons developed more sophisticated tastes. That and they grew up and started doing their own things in their own lives. Our trips to Burger King together ceased, but my solitary trips did not. On those evenings when I find myself on my own, it remains a simple pattern.

I walk that half mile, which gets me out and about. I order my inexpensive food. I sit and watch the folks around me. They are, for the most part, people who inhabit a very different rung on the socioeconomic ladder. Teens who can afford only the least expensive of meals. Recent immigrants of every hue and nation, usually out as a family. The elderly, sometimes with their caregivers. Homeless folk, sitting and muttering to themselves. And middle-aged men wearing rumpled and mismatched clothes, eating alone.

Which, I suppose, is where I fit into the demographic mix.

Burger King is industrial food, heavily processed and mechanized. The calories I consume there are those produced by big business—cheap calories that reflect the efficiencies of mass production. There are nearly eighteen thousand Burger King franchises that help Restaurant Brand International, the conglomerate that owns BK, through to an annual revenue of $1.2 billion.

Recently, Burger King rolled out their take on the latest in plant-based "meat": their Impossible Whopper, made from a peculiar process that simulates a beef patty through genetic modification. It's pretty tasty and rings all the mouthfeel bells of meat without the ecological impact. It's also remarkably successful—at least it has been at launch. It has driven sales, and it's part of a broader move on the part of inexpensive, mass-produced fast food to include plant-based options. KFC is now offering a giant bucket of "nuggets" that involve no chicken whatsoever, and Taco Bell (part of the same conglomerate that owns KFC) is now rolling out ersatz Mexican cuisine that is similarly devoid of meat.

None of this checks any of the traditional ecological boxes. This food isn't locally sourced. It isn't organic. It isn't produced using the agricultural methods of Indigenous peoples and honoring their lifeways. It's more than a little weird in a this-is-what-they-eat-in-a-dystopian-future, is-it-made-of-people kind of way.

But it's plant based. As the chapter on vegetarianism described, that means it produces an order of magnitude less carbon than meat and requires far less arable land. It is a substantial leap in the direction of making the most of our harsher world and ensuring that there's sufficient food to stave off the threat of famine.

As much as local and small, slow and resilient, traditional and organic are my preferences, and as much as I look at megacorporations with a healthy degree of skepticism, I can't dismiss economies of scale out of hand for reasons of ideology. Because the truth is that big business does some things very well. If it didn't, it wouldn't thrive.

Scale brings with it economies and efficiencies that locality does not. As resources dwindle and as the cheap energy density of fossil fuels and the simple abundance of a formerly verdant world diminish, we'll need those efficiencies. We'll need the fierce energies of progress that rise from enterprise.

The ancient lifeways that developed over the thousands of years of human history will still have their place, and they can contribute to our well-being and the survival of humanity. From among the learnings and wisdom of countless peoples and traditions, we have a rich smorgasbord of knowledge that speaks to how to endure all forms of climate and microclimate. But those ways of life rose as human cultures adapted to the world on which our species evolved. They are specific to a planet we will no longer inhabit. They also rise from a world that carried a far smaller population of human beings. It should be no soul's desire that our thriving in the future will be purchased only at the cost of gigadeaths among the poorest of us. Whatever gifts we have are a blessing, and all of the best paths of both scale and intimate community will be required. We don't want to think this. We'd like our own particular preference to be the one solution. It's a natural thing. I feel it.

We want to think that there is nothing we cannot do ourselves and that our way is the only way. This is a thought that I have often because I am human. It is also an aspect of my maleness, I think sometimes, a function of that heady XY chromosomal combination. I have a confidence, sometimes rational, sometimes a little less rational, in being able to accomplish just about anything I actually set my mind to. We have confidence in our gifts and trust in those abilities we have been given. It's necessary for us to thrive.

Write a novel? I can do that. Ride a motorcycle thirty miles in a blinding snowstorm even though any sane person would stay put? I never doubted myself for a moment. And fishtailing down a packed, snow-covered Beltway full of stressed Washingtonian commuters is its own kind of epic. Stupid, but epic. Install a ceiling fan in my kitchen even though I still had no real understanding of wiring and no carpentry skills to speak of? It's been twenty years, and the thing hasn't fallen on our heads yet. (Of course, I rarely turn it on and am careful not to touch it, but it sure does look nice there in the kitchen.)

But just as we turn to ourselves, we tend to doubt others. There is this peculiar unwillingness to rely on skills and knowledge that are not our own. We're taught, as individualists and consumers, that we're not supposed to. It's about us, and *our* perspective, and *our* way of understanding the world. The gifts and voices of the Other are ignored, diminished, or dismissed.

The apostle Paul describes a very different approach to life as he wrote his letter to the church in Rome. In the twelfth chapter of Romans, Paul is talking about how we live our lives. It's a shift from the tone of his whole letter up to this point, which has been mostly heady abstraction and elegant prose. But as the letter ends, Paul takes a step back and starts grounding all that theology in more of what it actually means to walk the Way of Christ. Here is what all of that actually looks like, Paul says.

We get to the opening verses of Romans 12, talking about the body as a deep metaphor for the nature of how we participate in Christ, with each of us contributing our gifts toward the good of the church. Our purpose as Christians is to serve the broader good, just as every part of the body serves the good of the whole. Some lead, some praise, some give, some show mercy and kindness. But all Christians have as their purpose the business of contributing to the joy of all, each part of the other, each serving in their own way as an agent of God's love. This is an important image for Paul because he presents exactly the same image of the church as the body of Christ for two whole chapters of another letter in 1 Corinthians 12 and 13.

But before we reach Romans 12:4–8, the apostle makes another statement about the body. He suggests that we need to present ourselves as "living sacrifices," giving ourselves over to the reason for which God created us. The purpose given to our lives is one that comes not from the pattern of this world but from the pattern of the love that is the Holy Spirit of the living God.

Throughout the letter to the Romans, Paul makes constant reference to the battle within us between the Spirit of God and our desire to conform ourselves to the world around us.

But how does that struggle play out in our day-to-day lives?

Paul himself appears to have been a remarkably capable and self-sufficient person. He was a scholar, trained in the languages of Greek and Hebrew. He was gifted in the art of persuasion, knowing how to use language and culture to change hearts and minds. He was a craftsman who knew how to work cloth and canvas, a skilled maker of tents and shelters who could find a living for himself anywhere in the ancient world. And he was resilient, able to endure hardships and challenges—shipwrecks and imprisonment and tortures—that might break others.

And yet it was this Paul, who could have been justifiably full of his own bad self, who argues that the Christian path requires us to need one another. We need those who are different, those who bring different gifts to the table than we can. The Christian path involves brothers and sisters each finding their own gifts while also learning mutual reliance.

When we let our lives be ruled by that desire for control over others, we might think we're livin' large and in charge. We don't need any knowledge beyond what we already hold. We, personally and in our own silos, are the ones who know and can do it all.

But Paul would argue that in the eyes of God, we are deluding ourselves. In trying to do and be everything while unconnected with the gifts of others, we no longer are defined by the gifts God has given us. We are, instead, controlled by the world that we seek to control. Our lives may seem empowered and free. But for all intents and purposes, we are bound.

All of this reminds me of how one particular community deals with their lives together in faith: the Amish. There's a term among the Amish: *Hochmut*, they call it, in a phlegmy, Germanic kind of way. *Hochmut* is pride. It's an orientation of the self toward the self. It's the certainty that you are the center of all things. *Hochmut*, as the Amish consider it, is the height of all human sin. It's also the thing that shatters communities and tears apart our ability to work together.

Their goal, in their lives together, is *Gelassenheit*—a German word again—in which a community sets aside ego and exists, mutually, in a gracious interwoven giftedness, where no one soul seeks to dominate the others. It's a path that creates remarkable resilience, a fabric of life that creates a robust endurance.

If we are to thrive as our path leads us into a new, harsh world, openness to the giftedness we find in difference will be needed. We may celebrate and develop our ability to "tend our own gardens," as Voltaire would have put it. We may rejoice in local robustness and our deepening connection to our community.

But we can also celebrate the human drive to enterprise, to strive for new discovery, to find the best path. We can be both global and local, approaching these challenges at both micro and macro levels. The easy binaries that we used to categorize our old world will no longer serve us in the new. To endure and to thrive, we will need to move away from either/or and embrace the both/and. We need everything humanity can bring to bear as we work to adapt to our new world.

We need the highest virtues of conservatism: its focus on thrift and resilience, on trusting and being trustworthy, on the value of personal striving and the deep love of kin, on a sense of duty and honor.

We need the highest virtues of progressivism: of openness to new paths, of being willing to move away from things that just don't work anymore, of a resistance to injustices committed against those who lack power.

We need the insights of folk wisdom: drawn from peoples who have struggled and strived for thousands of years, finding ways to live in countless different climates.

We need the vision of science and progress: that human ability to both discover the truth of the reality we live in and imagine paths that create new possibilities for life and growth.

To endure, we'll need to open ourselves to all of that and to those who are different, tapping every aspect of the human experience. We'll need all the gifts and every last reserve of human strength.

CHAPTER 19

LEARN FAITH, HOPE, AND RESILIENCE

This has all been a lot. Too much, in fact. Looking at the scale and timeline of climate change, things can feel more than a little overwhelming. Humanity is just so clumsy, so slow to act. This is particularly true when you dig around looking for some way to understand just how big the problem might be. Has this ever happened before, we ask?

Human beings need a sense of control over things, and having some prior event that can inform us about the why and how of a possible future really does help us frame our response. Has there ever been a point in the millions of years of human history where the massive and abrupt release of stored carbon occurred? Is there something we can look to that might give us some clues as to how long this will last and how we might survive it?

The answer is yes. Something similar to our current cascading climate crisis may have happened before in the history of our small world. It was a while ago—meaning between 299 and 252 million years ago, as the Permian era transitioned into the Triassic era. (Because our planet isn't seven thousand years old. And it isn't flat. And Elvis didn't fake the moon landing. And you don't need to drink only distilled water and pure grain alcohol to protect your precious bodily fluids from commie fluoridation.)

So quite a while ago.

That event is known most dryly as the P–Tr event, which doesn't sound like too much to worry about. It's a little more intimidating when we call it the Permian–Triassic extinction because the word *extinction* doesn't sound all that positive. That period of the history of our little world also has another, less formal name among archaeologists. It's called the "Great Dying."

Which is . . . well. Sweet Lord Jesus, that's grim.

The reason for the morbid name is simple: over a period of about twenty thousand years, nearly all life on earth died out. Ninety-five percent of marine species went extinct, and nearly 70 percent of terrestrial life perished. Ninety-seven percent of all living things died. The cause of this extinction, as best we understand it today, was a sudden global increase in temperatures. And by "sudden," we mean on a geologic scale, over thousands of years. The cause of that abrupt increase in temperature was most likely a combination of factors. A sustained period of volcanic eruption related to significant tectonic activity seems to have been a trigger event. But given that volcanism is generally connected to cooling events, something else must have been at play.

Those eruptions also seem to have triggered a massive burn-off of coal deposits in what is now Siberia—deposits that had amassed over millions of years of life on the young earth. This epochal burn-off of stored carbon increased atmospheric carbon dioxide levels explosively, trapping the heat of the sun and warming the seas. With sea temperature levels suddenly higher, the impact on early life was inescapable. The riot of invertebrate life that filled Panthalassa—the great, unified primeval ocean of the Permian period—couldn't adapt quickly enough. So those creatures did what all creatures do when they can't adapt to new realities. They died.

What may have adapted just fine were microorganisms that produced methane gas. As heat-trapping methane poured into the atmosphere, our planet grew even hotter. The terrestrial ecosystem that had evolved to thrive in the Permian climate crashed and failed. Over a period of between twenty thousand and two hundred thousand years, almost everything died. Our planet became a harsh world in which very little life remained.

Which, again, is why the P–Tr event is typically called the Great Dying. It is also one of the closest analogs to what is happening right now on our planet. The die-offs of marine life are just the leading edge of what is coming.

This is some harsh, terrible stuff. It is so huge and so harsh, in fact, that it does something to the human soul. The more we look at the scope and potential impact of climate change, the easier it becomes to give in to a sense of despair and fatalism. It's just so big. It's just so inescapably brutal.

Watching our global nonresponse to climate change, and with things trending in self-evidently terrible ways, there's been one consistent response. It's called "climate anxiety." As we encounter things beyond our control, we human beings get stressed out. We begin to ruminate, and overthink, and worry. That's true about our kids and our careers, about our relationships and our economy. But those things aren't anywhere near as vast and out of our personal control as an epochal shift of planetary ecology.

Faced with this seemingly insurmountable catastrophe, it's easy to feel lost and helpless. The deeper you get into it, the more lost and helpless you feel. The seas are rising! The crops are failing! Things are on fire! Everything is going to die! There's nothing we can do about it!

Globally, the impacts of this inexorable, inescapable crisis are weighing heavy on the souls of millions. As the weight of our planetary transformation becomes more evident, we get more depressed and panicky. We see and feel climate change in everything. If a person is already vulnerable and stressed, encountering the rising roar of our changing, hostile planet can prove to be too much. It's too complex. It defies our ability to change or escape it. The psychological impacts of this event are unquestionable: it messes with us. We become broken and listless, helpless and lost. There's a strong desire to curl into a fetal ball and wait for the rising seas to swallow us up.

That's where faith comes in.

During my time in seminary, I had the privilege of delving into some of the latest approaches to blending Christian faith with therapeutic counseling. Over and over again during my coursework, a theme was reinforced. When you're struggling with a mess or trying to rebuild a life, nothing is more potent and more valuable than hope. That's true in dealing with a crisis of relationship, or a vocational struggle, or even a struggle with mortality itself. It is the basis and bulwark of almost every intervention we were taught because without a sense of hopefulness, nothing gets anywhere.

Hope builds resilience. If you're going to survive something disastrous—something like an earthquake, a hurricane, or a sharknado—a sense that things will get better is absolutely key. It doesn't guarantee anything, but if there's one factor that makes it more likely that you're going to make it through a time of trial, it's your ability to hold out the possibility of getting through it.

In story after story of individuals who have survived and endured times of crisis and hardship, the one great unifying factor is that they never, ever let go of hope. Lost in the woods and injured, they keep striving. Alone in the vastness of the sea, they keep striving. Faced with a terrible prognosis, they nonetheless fight against that cancer. Imprisoned in concentration camps by demonic regimes, they keep striving because they've not lost sight of the possibility that more awaits them. They sneak northward, following the Underground Railroad in the cover of night, fleeing their enslavement by a hostile people who consider them less than human. They don't let go of the possibility that God is entirely capable of delivering them.

There are countless such stories of faith providing the strength to survive, but the one that rises to my soul as I affirm this truth comes from an old friend of my family. The Gulleys were Methodist missionaries whom my family befriended in Nigeria. Jim Gulley's life work was helping struggling communities with agricultural development. He had spent much of his life in some of the hardest hit and most impoverished countries of the earth, and in January 2010, that calling drew him to return with a group of missionaries to Haiti.

On January 12, 2010, Haiti was rocked by a massive earthquake, and the hotel in which Jim Gulley was staying collapsed around him. He and four other souls were trapped, all seriously injured, in a tiny space in utter darkness. For fifty-five hours, they lay in a five-foot by eight-foot by three-foot space. They prayed. They encouraged one another. They hoped together, though they were in pain and utter darkness, with no way of knowing how bad it was outside.

Two of the Methodist missionaries died, their injuries too great. But Jim survived, and when he had healed, he returned to Haiti to continue his work, a work that he considers his calling. In a January 2011 interview, Jim said, "I've had people ask me, well, didn't you find it hard to go back? And I had to say, 'No.' I didn't find it hard to go back. In fact, I would have found it hard not to go back because I was doing what I was called to do when I was there. And I could not abandon it. I needed to go back and to re-engage."

Faith has the power to help us work through and past our trauma and anxiety. It doesn't mean that those things aren't a real part of our experience. But it does mean that we overcome that sense of helplessness and act. The hope that rises from faith overcomes our fears, setting us fiercely toward what we are called to do each and every day.

And that, in story after story of human survival against the odds, is a great common thread. To survive when everything goes to heck in a handbasket involves two things. First, holding a deeply positive faith in the possibility of goodness; and second, acting, day by day and moment by moment, in simple ways, to move toward that reality. You trust in God, but you remember to collect the rainwater to drink. You remember your love for those who are close to you, and you attend to getting a little something to eat.

Again, hope builds resilience. Hope does not mean denying that there's a crisis. Hope does not mean living in a delusion constructed to avoid action. Hope simply looks beyond what is and toward the possible future in which the crisis is resolved.

Hope motivates and gives direction. Without a sense that things can ever get better, you're not going to seek help or take specific and concrete steps to live in ways that increase your chances of getting through a time of trial.

You can know this cognitively. You can hold those necessary steps in your mind and yet still struggle with hope. Or you can be like the people of Israel, who often found themselves in places where hope seemed completely pointless. One of the most powerful among the many stories of hope in the face of hopelessness comes from the book of the prophet Isaiah. It's from what scholars call Second Isaiah, the middle part of that great prophetic book, and that means that it's a story told when the people of Israel were stuck between a rock and a hard place.

It's hard for us, in the comfort of our homes, to really grasp how much the people of Israel had lost when Babylon destroyed Jerusalem and looted the temple. Having lived inside the Beltway my whole life, I suppose I might know that feeling if somehow Washington were to lay in ruins, with the National Mall destroyed, the Constitution ripped from its hardened shelter in the National Archives and burned, and the might of America's military and economy no longer anywhere in evidence. Just like the Judeans living in Jerusalem, I'd then be dragged off to a strange land to the north as a captive. (There the image breaks down a bit because for me as an American citizen, I'd be getting dragged off to Canada, where they'd probably just make me be more polite to people and give me access to health care. Oh, the horror.)

It's hard, as a people, for us to grasp how shattering that time would have been for the people of Judah. Hope, when you're in the deepest possible despair and everything you've worked toward is a shattered ruin? It feels impossible.

And what Isaiah was telling them was a wild and impossible thing. Hear it through those ancient ears if you can. You've been beaten down, broken, and trampled. Everything you thought was solid and real and eternal in the world had been smashed. Around you, a great and powerful empire dominates the world, and they've done a great job of destroying your identity as a people. They are everything that you hear and see, and you belong to them. You are nothing.

Isaiah acknowledges this. He names it. He's speaking to the "one deeply despised, abhorred by the nations" (Isaiah 49:7). He's talking to everyone around him. He's speaking to people who are enslaved, to those who see power all around them and know that they don't have any of it. Everything that was sacred and precious to them—everything that made them who they were—all of that will be forgotten.

The particular sin of Babylon, as an empire, was that it sought to utterly annihilate the identities of the people it enslaved. Unlike Assyria or Egypt, the goal of Babylon was to erase the heritage of all of the peoples they conquered. Left with no sense of their past and no vision for their future, those who were dragged in bondage to Babylon were meant to abandon all sense of their worth as persons. It was Babylon's curse on the ancient world, a curse that cuts far too close to America's own original sin of race-based chattel slavery.

To those people, Isaiah hits them up with a hope that is so far beyond their hoping that it must have seemed completely insane. Let's play that out a little bit. If you had sat down with one of those Judean slaves in Babylon and you had asked them, "What's the wildest hope you can imagine?" what would they have said? They'd probably have sat there for a while and struggled to come up with anything.

Maybe some more food would be nice. Or maybe fewer beatings? Fewer beatings would be a pleasant change.

If you gave it a little time and pressed them a little harder, they might have dreamed wildly that someday, somehow, they might be allowed to return to their land. Oh, it seemed completely crazy because there'd never been an empire as mighty as Babylon, with its wonders and gold and gardens and armies. But maybe we'll get to go back home to the life we knew. That would have seemed like a wild and distant fantasy, a Walter Mitty daydream going *pocketa pocketa* through your mind, almost totally divorced from reality.

But the message received by the prophet was that even this image they had in their minds wasn't enough. It couldn't even begin to match God's future for their people. Instead, to this broken people, Isaiah proclaims, speaking in the voice of the Lord, "I will also make You a light of the nations so that My salvation may reach to the end of the earth" (Isaiah 49:6 NASB).

This seems to go beyond hope and into the land of double-extra crazy, completely wild and impossible—hope that is so preposterously beyond the reality that the people of Israel knew that it might as well have involved alien crop circles or trying to actually summon a Patronus.

And yet here you are, reading this book thousands of years later. I am referencing exactly that prophecy. What do we have to say to those Judeans in their slavery? What does the reality of this country, this place, and your reading of these words have to say?

Imagine yourself suddenly back there, sitting with a displaced, dispossessed Judean by the rivers of Babylon. They look up at you and ask you, "What's going to become of us? Is there any future for the things that we hold sacred?" What could you honestly say to such a person?

You could honestly say that 2,500 years from now, there will be a nation that is utterly unknown to you. It will be a nation of many peoples, from many lands. They will speak a language that your world has never heard. It will be filled with strange and powerful magics. People will fly great metal ships in the sky. There will be tools that think and speak, pictures that seem to live and move. These people are so powerful that they make your Babylonian masters look like naive bumpkins. They can reach out their hands and touch the moon and the stars in the heavens.

Among this strange and magical people, hundreds of millions will still be reading these words, and the stories of your people will be sacred to them.

What sort of hope would that be? Your words would affirm Isaiah's wildly impossible proclamation, and it would be completely and utterly true. The reality of God's creation can spin out futures that not only meet our hopes but make our hopes look like a shadow by comparison. And within the sacred stories of faith lie hopes that span not months, not years, but millennia.

Therein lies the place for faith in this time of crisis. My faith remembers hopes that stretch across the span of millennia—hopes that have been proven true by the arc of human history. Deeper still, the deep magic of Christian faith asserts that there is a hope that transcends time itself.

That we feel a teensy bit stressed by this shift in epochs is understandable. That there's fear in our hearts is entirely human. We look to our future and see a terrible, rising storm, and we feel weak and small and very breakable. But hope overcomes fear, even the fears we feel deeply. The God to whom I turn in unbreakable hope is larger than the coming time of upheaval, greater than our pretty little blue dot of a world, infinitely larger than all of the time and space of our universe.

That's not to say things are going to be anything other than harder. The new and harsher world that is dawning is going to make demands of us that will require new depths of resilience and ingenuity. It will test our moral core, and it will require us human beings to find a deeper sense of purpose.

It will be a place of hardship. There will be bleak moments. Possibly bleak years or centuries.

But in times of hardship, the hope that rises from faith is deeply, powerfully adaptive. It's perhaps the most important gift we receive from following Jesus, and it is the most significant contribution of faith as humanity endures and overcomes.

It's from that hopefulness that I am certain we'll find the strength to act in our new and more demanding world. Every single moral action, every response, every moment of welcome or sacrifice? They will all rise from a deep certainty of the possibility that, with the grace of God, we will somehow manage to endure.

CONCLUSION

N one of this is going to be easy. Living with our newly angered Eden is, in fact, likely to be the most sustained and difficult time of hardship human beings have ever faced.

As humanity stares into the chasm of climate change, a deep paradox for Christians presents itself. On the one hand, this is an event unprecedented in human history. It presents us with challenges and threats that will put pressure on every aspect of human society. Every corner of the earth will be impacted.

On the other hand, the depth of the challenge posed by the climate crisis doesn't change any of the core moral obligations of the Christian tradition. Those essential virtues remain as relevant as ever. Teachings about wisdom and thrift, patience and resilience, welcome and mercy, grace and justice? All of them were essential to life before the climate crisis, and every last one of them is key to humanity thriving in the world that is to come.

The most challenging of all of Christ's teachings in this time of crisis will be love. Anxious, angry, frightened human beings are terrible at loving one another. We really are. Just as individual souls in times of stress can fall back into bad habits and destructive behaviors, so too do human cultures. The more we're pressed, the more we fall back into factionalism, recriminations, and all of the old hatreds that have made so much of human history a dismal, bloody mess. It will be equally easy to become paralyzed with fear at the scope and scale of this epochal event. A global calamity that significantly exceeds the scale of our lives can easily cause us to surrender to cynicism or anger.

For folks who claim to follow Jesus, there's this story he told about applying his moral teachings in times of crisis. It comes right at the end of the Sermon on the Mount, as Jesus pitched out yet another one of his pithy bits of soul-stretching wisdom. His listeners have just heard him present the most fundamental expectations of his way: humility, mercy, a nonanxious stance toward the world, and the radical love of all human beings. Now he's bringing it home for them.

The story is about how to build when you're facing a storm. We are presented with two men. The first builds his house on sand. The second builds on solid rock. The storm rises, and the one who builds his house on sand loses everything. But the one who has built on solid rock endures.

At the beginning of this book, I said that I harbor a healthy fear of creation in which we are all a small, fragile, mortal part. That remains true. Only fools approach the world in all its glory and terror with the assumption that it owes them anything.

But that is a mortal fear. Sure, I'm aware that I am a fleeting thing and that my life will fly forgotten as a dream at the opening day. Yet from faith I know this: all of us are so much more. We're grounded in a Love that transcends time and space itself. We're connected to God, who calls us to live that out, all the time, no matter what. We're taught by a Teacher and loved by a Friend who showed us what that means. We are redeemed and made whole when we embrace Christ's life as having authority over our own.

I've spent a lifetime following and straining to disciple my soul to the teachings of Jesus. What time I have is committed to the Way of Jesus and to sharing that Way with others. This little book has laid out what that means, right now, as the sweet Eden with which we were blessed grows hard and harsh around us.

It means slowing down and living wisely, being grounded in the great sacred tradition of biblical wisdom. It means consuming less and finding contentment in less. It means being prepared to let go of things we enjoy that matter less than the future of our children and our grandchildren—things like the delicious taste of a good steak, the power of a roaring engine, or the familiar feel of a favorite beach.

It means opening up our hearts and homes to those fleeing this crisis, treating them as we would want to be treated should we find ourselves as strangers in a strange land. It means setting aside the clumsy binaries of our pathologically adversarial society and looking with new eyes at the gifts of the Other. Where is the good that we should not forget? Where is the new thing that we need to discover?

And in all of that, hope. Dear Sweet Lord Jesus, it means hope.

It means being guided by hope even in the face of a crisis that we can mitigate but not avoid. It means seeing that there are real possibilities for humankind to make our way through this mess of our own making, particularly if we see this great unfolding change for what it is: something that demands our moral attention, our effort, and the best graces of our faith.

Creation is, after all, a wonderful and terrible thing. We need to take care of it, and of our souls, as we learn to survive in our angry Eden.

NINE WAYS FORWARD

So we've journeyed through the reality of climate change, established that it demands a Christian moral response, and laid out how the traditional virtues of the way of Jesus can be applied to this epochal crisis.

But you want more—not just general principles leavened with storytelling but specific things you can do to respond to this crisis. My natural inclination is to let folks figure out the best way to deal with things, but I do get that. Having a few clear instructions that we can operationalize is nice. We don't do well if we've got to figure it out on our own. "Love your neighbor," Jesus said and told us a story about a Samaritan. We listened to those stories and moral principles and promptly spent the last two thousand years killing, enslaving, and oppressing anyone who was different from us while simultaneously claiming Jesus was our Lord and Savior.

Well, maybe you haven't done that personally, but Lord knows we have together.

So in light of our human tendency to spin and skew things that are less concrete, here are nine more suggestions for how you can respond to the growing impacts of climate change.

1. Adapt to Diaspora

Holding on to our sense of self when we're forced to be constantly on the move can be hard, and it ain't gonna get easier as our warming world drives us from the coastlines and pushes humanity toward the poles. As we human beings are displaced from the places that give us a sense of who we are, how can we adapt? How do we prevent ourselves from losing a sense of connectedness to the past that shaped our present?

The answer to that is to remember where you're from. Not in the what's-my-street-address sort of way but instead by holding on to the complex narratives of your heritage. What are the peoples, places, and cultures that contributed to the reality of your existence? Know their stories, their myths, and their music.

Human beings need that sense of connectedness to heritage and story. The first congregation I served as a pastor was attempting to merge with a Korean American church. As part of that effort, we supported and provided resources to an organization that helped connect Korean adoptees to their cultural heritage.

One of the challenges that cross-cultural adoptees can face is a sense that they're missing a part of themselves. Even with the loving support of adoptive parents, that feeling of being out of place can gnaw at a soul. The ASIA Families Culture School provides Korean adoptees with a grounding in the culture of their birth, teaching language, art, music, and the basic elements of Korean heritage. With that knowledge, Korean American children in adoptive families can better grow and thrive.

For other Americans, delving into heritage has an additional benefit. We are—with the notable exception of Indigenous and First Nations folk—not from where we live. A quick dive into our genetic history through the miracle of modern testing lets us know that we're from all over the place. Bits and pieces of our DNA tag us as having not just a single heritage but multiple heritages. Rather than the clumsy erasure of identity that comes with modern-era racial categorization, DNA testing presents most Americans with a nuanced picture of their ancestry.

Our robust mongrel histories are complex, and being intentionally aware of the smorgasbord of bloodlines from which we spring reminds us that we have no more claim to the place we find ourselves than others do.

2. Hospitality

Welcome is an art that can be learned. We learn it by doing it: by opening ourselves up to the stranger, to the outcast, and to the refugee. Not by talking about it but by actually, physically doing it. That's a difficult choice in a fearful time. When things get rough, our natural human instinct is often to shut and bolt the doors, hold on to what we have, and hope that those whose voices cry for our help from outside simply go away. But that's precisely the opposite of what Jesus would have us do. What Jesus would have us do is welcome the stranger. Period.

When I was still a little boy, the congregation I currently pastor had a choice to make. The brutal war in Vietnam had ended, and South Vietnamese families were fleeing for their lives. They were risking it all, packed into boats out on the open Pacific. America, having had something of a hand in their plight, chose to open its doors to those who were then often called "boat people." My church was—and is—small and poor, but they had a manse on their property, one that would be suitable for housing a family of refugees for a time. The church was too small to support a full-time pastor, so no one was living there and the house was only partially used. There was available space. There was a need. The Spirit made the connection in the souls of the church members, and action was taken.

They connected with a refugee resettlement agency. They cobbled together funds to get the manse in livable order. And then they welcomed that Vietnamese family, helped them find their bearings, and cared for them until they were ready to find their way. That story was still part of the self-understanding of the community when I started serving them decades later.

Was it easy? Not really. Did it require sacrifice? Sure it did. But that is true of every good thing, or so Jesus-folk should know.

The best way to prepare ourselves for the coming crisis of climate refugees is to be doing that now. So assess your resources if you are in a congregation. Partner with other churches and faith communities, if need be. Find a refugee agency. Connect with people in need of welcome. Then do what you can with what you have.

Because if you're reading this, you likely have enough.

3. Slow Down, Find Sabbath

Slowing down is hard. It requires a sea change in your entire attitude about life. You have to be willing to do less, to take your time, and to let go of the desire to control every last moment of time. We are terrible at this. This is the heart of Sabbath: the fundamental commandment to just give it a rest already, people. In our Veruca Salt, I-want-it-now culture, engaging in Sabbath and living at a slower pace is an act of defiance.

It's a pretty chill act of defiance, but it requires intention. Sabbath isn't another thing to add to your to-do list. It's a complete shifting of the way you approach the life with which you have been blessed. It's a change in your ethos, that framework of self-understanding that guides all of your actions. It's a shift from anxiety to calmness of soul. It's a shift from a sense of lack and grasping to a sense of abundance and patient presence.

There's a peculiar irony in trying to fold that change into a listicle of stuff to do, but hey. Irony can be entertaining. There are two specific ways I find the spirit of Sabbath grows in a soul.

First, walk. I am not talking about the very specific spiritual practice of walking meditation. Walking meditation is lovely, and at different times in my life, I've used it to center myself. Should you want to do that, I'd recommend the teachings of Thich Nhat Hanh. But I'm suggesting here just taking a long walk. Several times a week, make a point of getting out of your SUV and doing something at a human pace. Walk to the store. Walk to the bank. Walk to the library. It will take longer. It will require you to be a little warm, or perhaps a little cold. The sun may get in your eyes. You can manage.

Leave your phone in your pocket. Or at home. Then walk. It can be a walk to the bank or a walk to get dog food. It can be a walk to the library or to a local restaurant. Or to church.

Walking changes your awareness of things. You get a better sense of the scale of the world around you and your place within it. And you're moving more slowly, and with intention, and that seeps into your sense of the proper pace of life.

Second, garden. Plant some seeds, care for them, and watch them grow. If you've got the room in your yard, set up a couple of raised beds. Have less space than that? A couple of modest, half-barrel planters will fit just about anywhere and can yield all kinds of delicious things. Live in an apartment? You can find a community garden nearby. Then get to planting. A tomato plant or three. Carrots. Beets. A little patch of strawberries. You name it, give it a go.

When gardens grow, they grow at their own pace and in their own time. You can help them along a bit. Good soil, the right amount of light and water, and they'll grow, but you've got to wait for them to do what they do. Gardens require work, but they also require patience.

The act of planting and tending to the growth of God's creatures reminds us of the limits of our control over things. It requires that we forget the ethic of instant gratification because it requires time, attention, and patience.

Both walking and gardening connect us with God's work around us. They both require us to slow down, to catch our breath, and to live at a pace that reflects our created nature. To slow into Sabbath, there's nothing quite like taking that walk in the garden in the cool of the day.

4. Live Humbly

Living humbly requires that change of mindset, a reorientation of your soul toward the modest, the simple, and the necessary. What do you actually need? What is genuinely required? What has value, and what does not?

There are all manner of ways you can change the lens with which you view the world and adjust what you see as having value. One I've found to be particularly efficacious is composting.

Soon after I started gardening, I decided to take a swing at turning "waste" into soil. There are all sorts of ways to do this. Rather than buying a compost barrel (which is excellent), I make my soil the old-fashioned way: a big pile. Into that pile I put all of the nitrogen-rich grass clippings from my yard and every carbon-rich leaf that falls in autumn. I also collect kitchen scraps in a little bin, which fills with coffee grounds and veggie leavings. All of that goes into the pile. I turn it with a pitchfork a couple of times a week or whenever I dump the kitchen bin. Between the plentiful worms and the aerobic bacteria, the pile turns to rich soil every six months or so.

Not just rich soil. Living soil. It's complex, it's organic, and when you turn it in the middle of the summer, it's warm. You can feel the heat rising from it as living things break down matter into the earth that will create more life.

That process changes how we see things. Everything has a purpose. Tops of overripe tomatoes. Carrot peelings. Coffee grounds. When I mow my lawn, it's not a chore. It's a harvest of nitrogen-rich greens. When the trees on my property start to blush with color in September, I don't see an annoying bother to be set out on the street for removal. I see a great bounty of life-giving carbon locked away in future earth, something of fundamental worth. Composting makes us aware of God's work in all of creation and of the impossibility of considering even the humblest leavings as anything other than good.

Claudio Gonzales is an elder at Casa Vidara, an intentional Christian community in Brazil. He describes the spiritual impact of composting:

> In the beginning, God created and said, "It is good." This banana peel that I am holding is good. And if God says it is good, then by whose authority do we treat it like garbage? The truth is, we are not authorized to treat it that way. And if you begin to handle your own food discards—like this

banana peel—as creation, instead of garbage, then it will never be garbage again. You will rethink what we are called to do with creation. So, man, it is subversive, because composting food waste is about learning to repent and rethink how we care for God's good creation.

Nothing teaches us the inherent value of the humble more than declaring every last thing good. If we can look to what we imagine to be "trash," worthy only of discarding, and see in it part of God's creative purpose? Our view of all things changes. And our desire to engage in the distracted hyperconsumption that is turning our world against us is diminished.

5. Be Vegetarian

A plant-based diet demands less of our small planet. It uses less land. It requires less energy to produce, process, and transport. Making the change to a vegetarian diet isn't particularly challenging. It really isn't. It isn't drab, and it isn't flavorless. It can be hearty, satisfying, and a little bit decadent on occasion. Making that transition requires a little bit of a change in mindset, but it has never been easier.

Start by folding a vegetarian meal into your diet each week. Make a hearty veggie lasagna with protein crumbles. Or a nice big batch of spaghetti, with some store-bought sauce bumped out with roasted peppers, mushrooms, olives, and black beans. You can eat off of those for a while, and the lasagna in particular only gets better with time.

You can go Mediterranean. Get some falafel mix, some good pita, and the appropriate slathering sauces. Hummus. Tzatziki. A block of feta. Sliced tomatoes, from your own garden if you can. Kalamata olives. It's quick and easy and tasty as all get-out.

Too pricey? There's nothing heartier, satisfying, and dirt cheap than the humble grilled cheese sandwich coupled with a tomato soup. Plus, that's not exactly hard to prepare.

Do that every once in a while. Then ask yourself, Hey, could I live off of this? You'll discover that you most likely could. Does that mean you will? Possibly. Possibly not. But coming to the realization that you can function just fine without meat is the first step in the process.

So give it a try. It's surprisingly delicious.

6. Render unto the Republic

I'm writing this in the early fall of 2020, and I'm honestly not sure how things will look when this book is published. The basic expectation of free and fair elections in America has been called into question for the first time in the lifetimes of any of us.

We're faced with an unprecedented array of crises. A global pandemic. A collapsed economy. And, of course, the spreading impacts of our hardening, harsher world. Fire tornadoes in the West. Rolling blackouts. Derechos devastating the corn harvest and shattering cities, as we saw in Iowa in 2020.

In the face of all of these existential threats, a substantial proportion of America has embraced a fundamental cynicism about our electoral system. Voting, or so would-be despots tell us, is corrupt.

The only reason this is being said, of course, is to heighten the skepticism that all authoritarians require to justify their rule. Things can't be any better, or so argue those who want us to believe that. "Voting doesn't mean anything," they whisper. "Don't even bother."

While this is a very human and very familiar response, it is also desperately wrong. Authoritarian systems that rely on misinformation for their power respond very poorly to events that challenge their assertions. It is always far easier for systems that are disconnected from the reality on the ground to pitch out propaganda, blame, and self-serving rationalizations instead of dealing with issues directly. In the absence of engaged citizens speaking up for their constitutional rights and ensuring that their government is responsive and on top of threats to our well-being, power will do whatever it wants.

The counterbalance to that is political engagement. Similarly, the best counterbalance to the ideologues and hucksters who deny climate change for their own benefit is political engagement. Getting out there and voting for leaders who take the climate crisis seriously is central to our adapting and enduring. Support the candidate who is most likely to bend the arc of history in the way it needs to go. It's your moral responsibility.

Do it.

7. Embody Grace

This one is hard. Faced with an existential crisis, one that jeopardizes our future and presents a real danger to our present, there are a large number of souls who choose to believe that the climate crisis isn't happening. I know some of them. The arguments are familiar.

It's fake. It's a hoax. What we're seeing is "just weather." Everything is normal, and this is just a thing that liberals, leftists, and the global elite are cooking up to justify taking away freedoms.

It is natural, in the face of that kind of stubborn intransigence, to feel anger.

Similarly, we inhabit a culture where attention to the climate crisis seems the farthest thing away from the minds of those around us. Even if there's not active denial, we need only look around our neighborhoods to see that America doesn't really care. Our suburban driveways are cluttered with SUVs and pickup trucks. The new homes being built are larger and more densely packed, using more energy and leaving less space for trees. Parents rush their kids to activities. Commuters while away their lives in traffic. Nothing changes.

It is, again, natural to feel anger at obliviousness. Anger is, after all, an entirely legitimate response to falsehood, injustice, and harm. It is an equally legitimate response to things that threaten us.

But the powerful energy that anger gives can itself cause harm if not leavened and guided by grace. Sure, we could give an earful to that relative who's convinced the whole thing is fake. Yes, we could give the stink eye to that soon-to-be-former friend who drives around in a great honkin' truck. None of that will be persuasive. When faced with the choice between discovering a new way of living and doubling down on what we already know, human beings will nearly always choose the latter. We can get louder, we can get angrier, we can shout, and not a dang thing will get through to someone whose heart has hardened. Not a thing.

Ultimately, the voice that will unplug the ears of those who won't act isn't ours. It's the voice of God's creation. When firestorms rage across the American West, and bow-shaped derecho clouds race across the heartland, and floods and storms drown the South? That will not be your voice speaking.

As much as folks who reject science and ignore the calls of the faithful might want to shut out that voice, they cannot block or unfriend creation. They can't listen to deniers on social media when they have no electricity. They can't just go about their business when roads are flooded and blocked with debris.

It takes more than the cries of a people to break the hardened heart of Pharaoh.

When that wall of denial begins to crumble, how will we respond? Will it be with a snarl, with "I told you so" and bitter rejection? Our task is to do what we know is God's will, follow God's most gracious path, and when others come to that realization, welcome them in. Not to judge, not to attack, not to assail but to encourage and witness in word and deed.

There's a second aspect of grace that's worth exploring here: the knowledge that we ourselves are imperfect. In some way, we're all responsible for this. Everyone who lives in industrialized societies has participated in or benefited from the overconsumption of fossil fuels that set us on this cascading path. That sense of guilt folds into our anxiety about human impacts on our planet, sometimes to the point of making people feel that it would be better if humankind didn't exist at all.

This sort of response is comprehensible but bitterly self-annihilating and dysfunctional. Sure, we've failed. But we can only do what we can do. We can choose to do the best we can and act in ways that mitigate our impacts.

The God that created us and that we know through Jesus of Nazareth understands our weakness and forgives. So, no, you're not perfect. But you're trying. Trust that this is enough.

8. Use Every Gift

Faced with the need to change our ways radically to reduce the production of atmospheric carbon, it feels like everything must change. How we get around. How we communicate. How we structure our day-to-day lives. How we eat and drink. Every facet of our existence seems woven into an economy that is utterly dependent on fossil fuels, and we're not quite sure where to begin. We are presented with an array of decisions that seem too many and too complex for us to embrace, and most of us aren't particularly well suited to transition to a hunter-gatherer existence. I know I'm not.

What we can do, though, is make sure that our choices—all of them, across the spread of our lives—reflect a commitment to both connect more deeply to our world and tread more lightly upon it. That means being guided by our awareness of the reality around us. Every choice, every action, every word, to the best of our ability. Does that mean we'll choose perfectly? No. But it does mean that we're endeavoring to act on our awareness in every sphere of our being.

Similarly, we can acknowledge and appreciate when other people and groups and corporations take action—even when those people and groups and corporations aren't ones we've seen as allies. Case in point: in the summer of 2020, the United States government did two desperately foolish things. First, it rescinded rules governing methane release and capture for fossil fuel companies. Methane, or natural gas, is a massive accelerant of the process of warming, and the more that is released into the atmosphere, the more rapidly the earth will heat. In the context of the climate crisis, it was an act of incomprehensible malicious harm on the part of the administration. One week later, the United States announced that, again by executive order, we would be selling the rights to explore for fossil fuels in the Arctic National Wildlife Refuge (ANWR). Not only did that choice threaten a pristine wilderness, but it also continued our dependence on fossil fuels in a way that could only be interpreted as malicious.

Resistance to these actions came from some interesting quarters. Fossil fuel companies, upon hearing that they could pad their profit margins by skirting methane capture regulations, said, "No. No, actually, release of methane is a problem for the planet, the regulations make sense, we've already adapted to that reality, and we'd rather not go back to doing it wrong."

In order to "develop" the resources that might lie under the ANWR, oil and gas companies would need to raise capital. That capital would need to come from investment banks, the big Wall Street players we're all so very fond of. Citigroup. Goldman Sachs. Those sorts of folks. To the great consternation of the administration, a consortium of the largest financial institutions in the world announced that they would not commit capital to any project in ANWR. To do so, they argued, would compromise the future of the planet. You can't do business on a ruined world, or so went their logic.

Massive financial institutions and oil companies don't generally pop to mind when I think of allies in the struggle to adapt to our warming planet. Yet as deeply as I prefer the small, the intimate, and the local, I'm aware that there are other allies in our efforts to endure this crisis.

Finding those places of connection and encouraging and honoring unexpected contributions to this struggle? That's a necessary part of any journey that we're all in together.

9. Learn Faith, Hope, and Resilience

The heart of all of this lies in our faith. Faith gives us purpose, and purpose guides our words and actions. As a Christian, I want every single thing I do to be measured against the teachings and life of Jesus. Am I doing what Jesus taught? Then he is my Lord and Savior. Am I ignoring him or spinning his words to justify my own desire for power? Then he is not my Lord and Savior, and any words of mine claiming to praise him are no more than straw in a refiner's fire.

Establishing faith requires more than just theologizing or complaining about "those people." You have to be personally, repeatedly transformed by Christ's call on your life. The way we do that is prayer.

If we're being honest, prayer may not be a strength for us. In the fading old-line denominations in particular, prayer tends to be deemphasized. It's too "personal." It's too impractical. We'd rather have meetings, or have meetings about meetings, or write earnest statements about how Something Must Be Done.

This is because we do not understand the primary function of prayer. It isn't a magical incantation that summons the power of our deity to do our bidding. God is already there by our side, doing all that is needed. Prayer creates in us a sense of connection to that purpose. It helps us define ourselves in terms of God's will and God's power, rather than the other way around.

Because prayer shapes and redefines us, it radically impacts our actions. When we pray for the well-being of others, our understanding of them changes. When we pray for guidance and strength, our understanding of ourselves changes. When we pray for intercession, our attention is opened to ways God's power is at work in the world around us.

There are many paths and techniques of prayer. Some are spoken. Some are silent. Some are physical. Some are sung. There are as many paths as need be for all of us to find our own way. If you've got one that has guided you and sustained you in grace, keep on keepin' on. If not, consider this option.

A discipline of daily prayer can be simple. Every day, I use the Lord's Prayer. I say it at waking. I say it during the day as it rises to hand. It's simple and unassuming and just about perfect. It's also the prayer that Jesus taught as the one prayer that's essential for every soul who follows him. Simple as it is, it is a reminder of my commitment to center myself on the moral guidance of Jesus in all things and to keep myself focused on the virtues that rise from his teachings.

It reminds me that my needs are truly simple and that the temptations of material wealth and power are illusions. It reminds me that I am connected to all people and all things.

Whatever grounds you in the Spirit, whatever reinforces Christ's purpose in your life? Use it.

ACKNOWLEDGMENTS

N o book springs to life without the inspiration, insights, and support of others, and there are folks out there whom I need to thank.

To the good souls of the National Capital Presbytery, thanks for challenging us to think about our connection to and responsibility for God's creation. Had you not issued that challenge, this book wouldn't exist.

To my stalwart agent Kathleen, thanks for your thoughtful, supportive reading and for believing in the manuscript. To Valerie, my appreciation for your energy, your encouragement, and your keen and gracious editorial eye. To all the souls at Broadleaf Books, thanks for giving the manuscript a voice.

To Joyce, thanks for your willingness to beta-read the manuscript, for your informed insights, and for gracing my social media feeds with reminders of the awesomeness of electric cars.

To the good souls of Poolesville Presbyterian Church, thanks for your support of my writing, for your sustentation of my soul, and for all the ways y'all express what it means to live in faithful community together.

To Rache, for your support and enthusiasm.

And finally, to our warming world. Hey. We hear you. Please don't squish us. We'll do better.

NOTES

Chapter 2

13   ***"climate change is real, it's happening":*** "An Information Statement of the American Meteorological Society," American Meteorological Society, April 2019, https://tinyurl.com/y4parzbq.

13   ***"two-thirds of US strategic":*** Office of the Under Secretary of
Defense for Acquisition and Sustainment, *Report on Effects of a
Changing Climate to the Department of Defense*, January 2019,
https://tinyurl.com/y5ght82r.

14   ***"relevant to point out that":*** Office of the Under Secretary,
16–17.

14   ***"Both Shell and Exxon raised":*** Greenhouse Effect Working
Group, *The Greenhouse Effect*, May 1988, https://tinyurl.com/
yb88h4e5; M. B. Glaser, *CO2 "Greenhouse" Effect*, November
1982, https://tinyurl.com/y5wgclcv.

14   ***"Catholic Church claims as members":*** "Presentation of the Pon-
tifical Yearbook 2019 and the Annuarium Statisticum Ecclesiae
2017," Holy See Press Office, June 3, 2019, https://tinyurl.com/
tjh56rf.

15   ***"the church officially acknowledged":*** *Laudato Si'* I:23–25.

16   ***"When the Lion had first":*** C. S. Lewis, *The Magician's Nephew*
(New York: Macmillan, 1970), 125–26.

Chapter 3

22 *"when the Union of Concerned Scientists"*: "1992 World Scientists' Warning to Humanity," Union of Concerned Scientists, July 16, 1992, updated October 29, 2002, https://tinyurl.com/y79n5l4e.

Chapter 4

30 *"learn a new word"*: Stephen F. Corfidi, Jeffry S. Evans, and Robert H. Johns, "About Derechos," National Oceanographic and Atmospheric Administration, May 2018, https://tinyurl.com/yxr6k6kp.

Chapter 5

37 *"that entire metric changes"*: John Wihbey, "Fly or Drive: Parsing the Evolving Climate Math," Yale Climate Connections, September 2015, https://tinyurl.com/m5vw7bj.

Chapter 7

47 *"time of energy abundance"*: John Browne and Jean-Marc Ollagnier, "We Live in an Age of Energy Abundance," *Forbes*, March 2013, https://tinyurl.com/yykxvvlw; Ken Cohen, "America's Energy Abundance Staving off a '70s-Style Crisis," ExxonMobil Perspectives, July 2014, https://tinyurl.com/y2fj29wp.

48 *"fossil fuel industry profits"*: Jeff Barron, "2018 Was Likely the Most Profitable Year for U.S. Oil Producers since 2013," US Energy Information Administration, May 2019, https://tinyurl.com/yxao97yl.

48   ***"ExxonMobil alone, in 2018":*** "ExxonMobil Earns $20.8 Billion in 2018; $6 Billion in Fourth Quarter," ExxonMobil, February 2019, https://tinyurl.com/y49n64yp.

48   ***"world's most profitable company":*** Stanley Reed, "Saudi Aramco Is World's Most Profitable Company, Beating Apple by Far," *New York Times*, April 1, 2019, https://tinyurl.com/y5ggp6nm.

48   ***"oil company Rosneft saw":*** Oliver Griffin, "Rosneft Posts Profit Fall on Revenue Drop, Forex," MarketWatch, February 5, 2019, https://tinyurl.com/y35ppjrx.

Chapter 8

53   ***"laid out the scenarios":*** David Wallace-Wells, "The Uninhabitable Earth," *New York Intelligencer*, July 2017, https://tinyurl.com/ybmjvwrj.

54   ***"wet-bulb temperatures":*** Steven Sherwood and Matthew Huber, "An Adaptability Limit to Climate Change Due to Heat Stress," *Proceedings of the National Academy of Sciences*, May 25, 2010, https://tinyurl.com/y6lsbs55.

54   ***"trying to feed more people":*** David Battisti and Rosamond Naylor, "Historical Warnings of Future Food Insecurity with Unprecedented Seasonal Heat," *Science*, January 9, 2009, https://tinyurl.com/yxjel3vg.

54   ***"rolling death smogs":*** Wallace-Wells, "Uninhabitable Earth."

54   ***"decrease in air quality":*** Julie Chao, "Elevated Indoor Carbon Dioxide Impairs Decision-Making Performance," Lawrence Berkeley National Laboratory News Center, October 12, 2012, https://tinyurl.com/yaa9uh65; Robinson Meyer, "How Climate Change Covered China in Smog," *Atlantic*, March 21, 2017, https://tinyurl.com/y5kt6umo.

55   ***"would ding our economy":*** M. Burke, S. Hsiang, and E. Miguel, "Global Non-linear Effect of Temperature on Economic

Production," *Stanford Nature* 527, no. 7577 (2015): 235–39, https://tinyurl.com/y9n3ytqk.

55 ***"rising sea levels, decreasing crop":*** M. Burke, S. Hsiang, and E. Miguel, "Climate and Conflict," *Annual Review of Economics* 7 (2015): 577–617, https://tinyurl.com/y2xw66z8.

55 ***"cycle continues to progress":*** The Permian–Triassic extinction, 252 million years ago, was a complex event that involved the collapse of the carbon cycle. It was worse than the Chicxulub extinction. Ninety-seven percent of species went extinct. More on this fun and delightful event's relevance to our current climate challenges in a later chapter!

57 ***"slowly, surely, declining":*** Michael Gerson, "Thinking about the Apocalypse," *Courier Journal*, April 7, 2015, https://tinyurl.com/yxd5oe3d.

Chapter 9

61 ***"lust in my heart":*** Bo Emerson, "When Jimmy Carter Lusted in His Heart," *Atlanta Journal-Constitution*, September 28, 2017, https://tinyurl.com/y5ho6rvq.

62 ***"Instead of asking yourself":*** George Macdonald, *Unspoken Sermons Series I, II, and III* (New York: Simon & Schuster, 2012), 212.

Chapter 10

73 ***"an epistle of straw":*** Martin Luther, *Luther's Works*, ed. E. Theodore Bachmann, vol. 35, *Word and Sacrament I* (Philadelphia: Fortress, 1960), 362.

Chapter 11

81 ***"span of a single generation":*** Robin Wall Kimmerer, *Braiding Sweetgrass: Indigenous Wisdom, Scientific Knowledge, and the Teachings of Plants* (Minneapolis: Milkweed Editions, 2013), 13.

Chapter 12

90 ***"word used to identify":*** This Yoruba/Igbo/pidgin word is of uncertain origin, and variant pronunciation depends on regional dialect. I'd had it explained to me as "person from another place" or "brother from afar," but it could also mean "person whose flesh has been cut away."

90 ***"has taken deep root":*** David McClendon, "Sub-Saharan Africa Will Be Home to Growing Shares of the World's Christians and Muslims," PRC FactTank, April 19, 2017, https://tinyurl.com/y4yhrnr5.

92 ***"played by Fred Wesley":*** If you do not know Fred Wesley, you should. Google "Pass the Peas Fred Wesley." Then listen and imagine how that'd change the feel of passing the peace in worship. You're welcome.

95 ***"In the Tanakh, we":*** This is scholarly shorthand for the Torah, Nebiim, and Ketubim (the Law, Prophets, and Writings), often called the Old Testament.

96 ***"Christ is neither taught":*** Luther, *Luther's Works*, 399.

97 ***"In the Talmud, rabbinic":*** Babylonian Talmud, *Bava M'tzia* 59b.

Chapter 13

105 ***"cripple the development of":*** Hara Marano, "A Nation of Wimps," *Psychology Today*, November 1, 2004, https://tinyurl.com/yxtabkp9.

110 ***"Germans work only":*** OECD, "Hours Worked," OECD Data, accessed November 17, 2020, https://data.oecd.org/emp/hours -worked.htm.

117 ***"food was our calling card":*** Edie Gross, "A Network of Black Farmers and Black Churches Delivers Fresh Food from Soil to Sanctuary," Faith and Leadership, May 28, 2019, https://tinyurl .com/yxd4axdw.

Chapter 14

125 ***"agricultural production losses":*** Hong Xu, Tracy E. Twine, and Evan Girvetz, "Climate Change and Maize Yield in Iowa," *PLOS One* 11, no. 5 (2016), https://tinyurl.com/yxhuzttd.

Chapter 19

167 ***"impacts of this inexorable":*** Victoria Knight, "Feeling Anxious about Climate Change? Therapists Say You're Not Alone," People .com, July 15, 2019, https://tinyurl.com/y4gfavry.

167 ***"weight of our planetary":*** Thomas J. Doherty and Susan Clayton, "The Psychological Impacts of Global Climate Change," *American Psychologist* 66, no. 4 (May–June 2011), https://tinyurl .com/y6h7nwyz.

169 ***"I've had people ask me":*** "Haiti Quake Survivor: Jim Gulley," People of the United Methodist Church, January 12, 2015, https:// tinyurl.com/y4ywsopm.

Nine Ways Forward

184 ***"In the beginning, God":*** Quoted in Sam Ewell, "Caring for Our Common Home," *Plough*, July 20, 2020, https://tinyurl.com/ yxt45ck2.